Making Sense of
Fractions, Ratios,
and Proportions

2002 Yearbook

Making Sense of
Fractions, Ratios,
and Proportions

2002 Yearbook

BONNIE LITWILLER
2002 YEARBOOK EDITOR
UNIVERSITY OF NORTHERN IOWA
CEDAR FALLS, IOWA

GEORGE BRIGHT
GENERAL YEARBOOK EDITOR
UNIVERSITY OF NORTH
CAROLINA AT GREENSBORO
GREENSBORO, NORTH CAROLINA

NATIONAL COUNCIL OF TEACHERS OF MATHEMATICS
RESTON, VIRGINIA

Library of Congress Cataloging-in-Publication Data:

ISBN 0-87353-519-7

Printed in the United States of America

Contents

Part 1: Elementary School

Part 2: Middle School

Preface

THE publication of the 2002 Yearbook, *Making Sense of Fractions, Ratios, and Proportions* and its accompanying Classroom Companion booklet marks a first for NCTM. This is the first yearbook to have a supplementary booklet that provides activities for teachers to use in their classrooms.

Fractions, ratios, and proportions are essential concepts in the middle school curriculum, but their development and understandings begin in the elementary school. This yearbook contains articles that give insights into students' thinking about these topics. Authors offer suggestions on how to develop the concepts and skills associated with fractions, ratios, and proportions.

The transition from additive to multiplicative reasoning is an important part of students' mathematical development. Teachers and their students need to be aware of the differences between additive and multiplicative reasoning and the types of problems that generate each type of reasoning. Authors provide suggestions to help both the teacher and his or her students in this transition.

Mathematics is a language, and that language plays an important role in making sense of fractions, ratios, and proportions. Phrases such as "50 percent larger than" and 50 percent as large as" have different meanings, but are often misused. Authors have been careful to use correct mathematical language. The teacher is a role model in the mathematics classroom and must be aware of the abuses and correct uses of language.

Proportional reasoning is fundamental to the understanding of the mathematics of the secondary school curriculum. Thus, it is important that middle school students have the developed ability to use the proportional reasoning process as they go on to more-advanced mathematics. The authors have given guidance in this area.

The division of the three parts of the yearbook is arbitrary by the panel. We divided the yearbook into elementary, middle school (junior high), and professional development sections. We have limited the yearbook to elementary and middle school where the curriculum stresses the development of understanding concepts, the development of computational algorithms, and the development of multiplicative and proportional reasoning. If students can make sense of fractions, ratios, and proportions and can reason in a logical way, the transition to secondary school curricula will be a smooth one.

The Classroom Challenge sections contain nontraditional problems for the teachers and his or her students. They can be given to one student, students working in pairs or groups, or to the class as a whole. These types of problems can be the basis of discussion for the evaluation of students' understanding of given concepts.

Many manuscripts were submitted to the Editorial Panel for review. I would like to express my appreciation to all the authors, especially those who have shared their expertise in this yearbook.

During the past three years, I have had the pleasure of working with outstanding mathematics educators, whose expertise and dedication have been without equal. My thanks to the 2002 Yearbook Editorial Panel:

> George Bright, *University of North Carolina at Greensboro, Greensboro, North Carolina*
>
> David Duncan, *University of Northern Iowa, Cedar Falls, Iowa*
>
> Judith Sowder, *San Diego State University, San Diego, California.*
>
> Teresa Yazujican, *Eureka Middle School, Eureka, Illinois.*

I am indebted to the Yearbook General Editor, George Bright, whose support and guidance kept us all on task. Thanks to the editorial and production staff at NCTM, who transferred the manuscripts into this quality yearbook. Thanks also to Nadine Lilleskov, the mathematics department secretary at the University of Northern Iowa, and to Tanya Scott, my graduate student, who gave me countless hours of help.

The authors of the articles in this yearbook devoted many hours to creating manuscripts that would inform readers and motivate them to teach mathematics to all students. We hope that the content of this yearbook will help the readers both teach mathematics better and teach better mathematics.

Bonnie Litwiller
2002 Yearbook Editor

1

Introduction

Judith Sowder

for the Editorial Panel

A THEME of the chapters in this section, and indeed throughout the entire book, is that of making sense of mathematics. Resnick (1986) once said that children who are successful in mathematics are those who believe that they can make sense of mathematics and put forth the effort to do so. Unfortunately, too many children, by the time they reach upper grades, have lost this belief just at a time when they most need it—when they begin to work with fractions and proportions. For those who have come to believe that mathematics is a set of rules to memorize and apply, this mathematics is indeed very difficult for them to learn.

A colleague of mine recently gave his class of teachers an assignment to look for an example of the "hidden" curriculum that they teach—that is, what they teach without even being aware they do so. A third-grade teacher came to class the following week and said that she realized that her students never asked "why" questions in mathematics class, although they regularly did so in other classes. She asked one of her better students why this was so. He replied that in mathematics there was nothing to ask "why" questions about—it was just a matter of applying the rules! This aspect of the teacher's hidden curriculum came as a revelation to her, particularly as she herself was becoming more mathematically confident in her university class and finding lots of "why" questions to ask her instructor.

The chapters here deal with ideas that are very difficult for many children but need not be. If solid groundwork is undertaken in the early elementary grades, then concepts related to fractions, ratios, and proportions take root and grow. Some of the ideas and terms found in these chapters, such as unitizing, additive reasoning, and multiplicative reasoning, may be new to some teachers, but these chapters will help clarify the terms and show why the concepts they refer to are important for teachers to understand. I am reminded of a fifth-grade teacher I visited who had decided that his students needed to learn to distinguish between situations for which additive reasoning was appropriate and those for which multiplicative reasoning was appropriate. He told his students a story about investing $2 in Nicole's bank and getting $8 back at the end of a year, whereas Phi invested $6 in Tabitha's bank and got $12 back at the end of a year. He asked who got the better deal, and his students voted overwhelmingly that the two deals were the same because they both got $6 back—an example of additive reasoning. They did not see that the first investment had quadrupled, whereas the second investment only doubled. The teacher then changed the numbers: he invested $1 and got $20 back in Nicole's bank after one year, and Phi invested $60 and got $79 back in Tabitha's bank after one year. Once again, the students insisted they both got the same deal: they each received $19 back. But then Henry said, "I'd put a dollar in Nicole's bank, because if I put a dollar in there, I'd get 19 back; if I keep on putting more dollars in there, it'll go past $79." Henry was beginning to reason multiplicatively. He provided the foundation for this teacher's next lessons.

Students can help us teach! Many of these chapters tell a similar story. All will give you much to think about as you plan your instruction.

Happy reading!

REFERENCE

Resnick, Lauren B. "The Development of Mathematical Intuition." In *Perspectives on Intellectual Development: The Minnesota Symposia on Child Psychology*, vol. 19, edited by Marion Perlmutter, pp. 159–94. Hillsdale, N.J.: Lawrence Erlbaum Associates, 1986.

2

The Development of Students' Knowledge of Fractions and Ratios

John P. Smith III

No AREA of elementary school mathematics is as mathematically rich, cognitively complicated, and difficult to teach as fractions, ratios, and proportionality. These ideas all express mathematical relationships: fractions and ratios are "relational" numbers. They are the first place in which students encounter numerals like "3/4" that represent relationships between two discrete or continuous quantities, rather than a single discrete ("three apples") or continuous quantity ("4 inches of rope"). Proportionality concerns an identity between two (or more) such relationships.

Students' experience with these concepts begins early—even before formal schooling—and extends well into the high school years. Before they learn anything in our classrooms, students engage in activities in the everyday world where they generate ideas about fractions, ratios, and proportionality. Some activities take place in the home (e.g., how to divide up an object among friends), some in organized practices in the wider culture (e.g., tracking baseball batting averages). In these activities children construct knowledge about relational numbers. They bring this *constructed* knowledge into our classrooms where it interacts with what our curriculum and teaching offer *(instructed knowledge)*. Mathematically successful students manage to connect these two bodies of knowledge; students who never really understand do not. It is our task as teachers to help students connect their constructed knowledge to the powerful new ideas we want to teach them. We must help students make sense of expressions like "3/5" and "$x/3 = 8/12$" in ways that (1) connect to their ideas and (2) address honestly the mathematics of rational numbers. But to do so, we need to listen for their ideas, which are often quite different from what we are teaching and are sometimes only partially correct. My main purpose in this article is to provide some guidance in what to listen for.

Given the complexity of people's experiences and ideas and of the mathematics of rational numbers, one short article cannot present a complete survey of the nature of students' ideas and how they arise and interact with instruction. The whole story is simply too complex. What this chapter offers is a reasonable initial answer to the question, "Where do students' ideas about fractions and ratios come from, and how can we work productively with them in the classroom?" Because more is known about their development, I shall focus more on the origins of fraction ideas than ratios.

Fractions, Ratios, Proportions, and Rational Numbers

To communicate clearly we must be careful about our terminology in this area of mathematics. Mathematics educators often talk right past one another because they are using *fraction, ratio,* or *proportion* in different ways without realizing it. My terminology may not strike you as either necessary or optimal, but I want to be clear about my basic terms to avoid miscommunication.

We must first acknowledge that when someone writes a numeral like "3/4," we don't know how they are thinking until they tell us. Indeed, *fraction* is such a powerful mathematical idea because it can be used to express so many different kinds of relationships. But the multiplicity of meanings can wreak havoc with our communication. For example, we typically use the same numeral to express both fractions and ratios, yet these ideas are quite different.

To provide some clarity on this point, I will use the term *quotient* to refer to numerals like "3/4" and "11/7" whenever the meaning of the numeral is ambiguous. When the context makes clear that we are referring to a divided quantity, I will use the term *fraction. Divided quantity* is a synonym for *partitioned quantity;* both terms refer to a whole quantity that has been divided into some number of equal-sized parts. When a quotient refers to a multiplicative relationship between two quantities (e.g., how many times larger or smaller one is than the other), I will use the term *ratio.* For example, the numeral "3/4" also expresses the relationship between the 12 boys and 16 girls in a classroom of 28 students (i.e., for every 3 boys, there are 4 girls). The word *multiplicative* is essential here because we could also describe the relationship between boys and girls in additive terms, that is, as a difference, "4 more girls."

Similarly, I will use *proportion* and *proportionality* to refer to reasoning with ratios. If ratios are multiplicative relationships between two quantities (as above), then proportional thinking involves projecting the same ratio into another situation (or situations). Writing equations like "3/4 = 6/8" and "2/3 = x/12" does not necessarily indicate proportional reasoning. In both cases, the person may be thinking about equivalent fractions. As with numerals, equations do not immediately reveal how the people who wrote them are thinking; the writer needs to explain.

All teachers of mathematics have very likely encountered the term *rational numbers*, in their education and teaching. How exactly do rational numbers relate to quotients, fractions, and ratios? In a word, the answer is "Not in any simple way." From the standard mathematical perspective, rational numbers are quotients; they take the form "a/b, where a and b are natural numbers, and $b \neq 0$." One way to think about rational numbers is that they are all the possible solutions to equations of the form, "$ax = b$," where the conditions above apply. Rational numbers have many important properties as a number system. They are dense; they make up an ordered field; they are closed with respect to both addition and multiplication; you can "build" the real numbers with them, to name just a few.

But how do rational numbers relate to fractions and ratios? Here there are important puzzles. For example, we can add ratios but we don't add them the way we add fractions. If "2/3" and "2/3" are ratios, then "2/3 + 2/3 = 4/6 = 2/3." And this makes perfect sense—as ratios. If Samantha got 2 hits out of 3 at-bats at both Tuesday's and Friday's baseball games, then her batting average for those two games is (2 + 2)/ (3 + 3) or "4/6." But as fractions, the same sum is "4/3" or "1 1/3." Instead of trying to resolve these puzzles, I would rather suggest a way to think about the connection among these three kinds of numbers. Rational numbers are a powerful mathematical abstraction from fractions and ratios, as we use them in our daily lives. Some properties of ratios, like how we add them, just don't fit the formal properties of rational numbers. If this sounds confusing, rest assured that it is not a simple matter. But in your teaching, a simple principle follows: Work with your students to understand fractions and ratios in their own terms first. If they don't grasp those ideas, the formal concept of rational numbers will make little sense later on.

THE ORIGINS OF FRACTIONS: PARTITIONING AND DIVIDED QUANTITY

As parents and caregivers of young children know, the need for fractions and the development of action sequences to generate them arise quite early in children's social activities with physical objects. Often objects (like cookies) are desirable and scarce and therefore must be divided up and shared. "Fair" sharing leads quickly to the necessity of parts of equal size. Piaget and his colleagues were among the first to explore the development of children's ability to partition simple objects (circular, square, or rectangular)—that is, how they make equal-sized parts for any given number (Piaget, Inhelder, and Szeminska 1960, especially chapter 12).

Piaget's task involved sharing a make-believe "cake" equally among some number of dolls (each representing another person in the situation). To solve this problem, children had to satisfy three conditions simultaneously: (1) Some part of the cake must be allotted to each doll, (2) the parts must be of equal size, and (3) the parts, taken together, must exhaust the whole. Preschool children, ages 4 and 5, sharing among three dolls would typically satisfy two conditions but violate the third in the process. For example, they would create 3 approximately equal parts but not exhaust the whole. Piaget's most important finding from this work was a developmental sequence of solutions. Children first construct the 2-partition (that is, equal shares for two people) and succeed at this quite early. Soon they can extend their solution to the 4-partition. Then after substantial struggle they can partition into 3 and its multiples and finally into 5 parts and other prime numbers. This sequence is sensible (a 4-partition should be easier than a 3-partition, since "halving" is early achievement), and has been confirmed in later research.

An important cognitive issue in children's partitioning is the interplay between constructing the partition (dividing the object itself or making a pencil drawing on paper) and holding an image of the partition in mind. Both are acts of mental construction, but only one is visible! For small numbers, children who cannot construct the partition may not be able to imagine it either. But for larger numbers, elementary-school-age children often have an image of the correct partition, but they struggle to recreate it on paper—for example, they cannot quite get the parts to come out equal.

Because children's drawings and mental images are linked, teachers can support their mutual development by providing opportunities for both drawing and correcting partitions and imagining them (e.g., by watching teachers build and adjust partitions or by working with flexible materials themselves). The use of manipulative materials, like Fractions Tiles (Jenkins and McLean 1972) and fraction strips, is important here; but we should not quickly link children's representations with such prepartitioned materials to their ability to construct and reason about their own work. As Ball (1992) has pointed out more generally, students' work with manipulative materials does not automatically generate mathematical knowledge. Fraction manipulatives, skillfully used by teachers, can support children's developing knowledge of fractions, but only when these materials are seen as representations of many, many examples of divided quantities.

Indeed middle grades teachers (grades 5–9) may be lucky enough to have students who can already partition (mentally and physically) into any given number of parts. If so, they can work with those students to develop meanings for fractions from divided quantities. But the prudent course is to check and see what partitions students can handle. Many children who cannot partition relatively easily will struggle with work that relates fraction numerals to divided quantities. Such students would profit much more from additional work on partitioning.

THE DEVELOPMENT OF STUDENTS' REASONING ABOUT FRACTIONS

Speaking very generally, children's knowledge of fractions moves through two broad phases of development: (1) making meaning for fractions by linking quotients to divided quantities and (2) exploring the mathematical properties of fractions as numbers. Indeed the "strange" properties of fractions that often make them so troublesome to learn (e.g., their "unnatural" properties of order and equivalence) can also make them engaging for students. If we include arithmetic operations in "properties"—as we will in this article—this chunk of mathematics represents one of the main instructional targets of the middle years (grades 5–9). It is difficult to teach this content for understanding; students "have it" one minute and lose it the next. A deep appreciation of the ideas and forms of reasoning that students bring to learn fractions will not solve all our instructional problems, but it will help to address them.

Starting with Fractions: Using Quotients to Name Divided Quantities

In some sense, helping students see how to use quotients to show and talk about fractions is not complex or difficult. Once students can partition, they can learn relatively quickly to work with, and talk about, collections of parts in those partitions. This learning is mostly a matter of students acquiring the accepted language for expressing what they are already doing. Once they appreciate the need to talk about collections of parts (that is, "three fourths" not just "one of the fourths"), the path to fractions is clear: The quotient, "3/4," represents a collection of 3 parts from an object divided into 4 parts. In conceptual development, it matters little whether we require students to learn the official terms *numerator* and *denominator* right away or temporarily accept "the top number" and "the bottom number." What matters is whether students understand that the meaning of the two numerical components is given by their position.

That said, numerous conceptual hurdles stand between students and a solid understanding of fractions as relational numbers. First, students need to grasp the key idea that fractions name the relationship between the collection of parts and the whole, not the size of the whole or its parts. For example, in figure 2.1 below, which fractional part is greater?

The best answer is "It depends." In total area, the "fourth" is greater. But if we interpret the question as asking for the greater relative amount, the correct answer is the "third." Students who are just learning fractions will need help seeing that they are looking for the relative, not absolute size or amount. Extensive activity in partitioning wholes in many different ways with many different sized objects is useful here. Care should be given to the language used to describe the relative sizes of parts, collec-

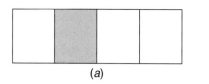

(a)   (b)

Fig. 2.1. 1/4 of a larger whole (a) versus 1/3 of smaller whole (b)

tions of parts, and wholes. For example, references to "big" and "bigger" in relation to divided quantities like those in figures 2.1 and 2.2 can be very confusing for students. All students should be able to explain in clear and convincing terms why the divided quantities in figure 2.2 do or do not represent 2/3 of the whole.

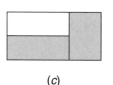

(a)   (b)   (c)   (d)

Fig. 2.2. An example (a) and nonexamples (b, c, d) of 2/3 of the whole

Students also need to become comfortable reasoning and talking about parts of discrete quantities (that is, collections of objects) as fractions. It makes just as much sense to talk about having "three-fourths of a sack of marbles" as it does to talk about "three-fourths of a cookie." Building in work with discrete and continuous quantities helps force the important task of identifying the unit (What is "one whole object" or "one whole set?"). Later on, when students are trying to make sense of arithmetic with fractions, they will need to keep the referent whole in mind.

Third, it is important to wrap language around all work with collections of parts in divided quantities. As students learn to use fractions (words and numbers) to talk relative amounts in divided quantities, they need reminders and support to be clear about what sort of part it is (e.g., "fifths" or "fourths"), how many there are, and what the referent whole is. As Thompson (1994) has pointed out, drawings do not speak for themselves. The same diagram of a divided quantity can represent many different relationships. In addition to the "standard" interpretation of figure 2.3 below as "3/5 of the whole is shaded," other sensible interpretations include: "1 1/2" (if the whole is the two unshaded parts), "3/2" (for the ratio of shaded to unshaded parts), and "2/3" (for the inverse ratio).

Patient work with students to develop clear and consistent language for divided quantities will resolve some common misconceptions about fractions (e.g., that "2/3" means "two objects, both divided into 3 equal parts" and that "5/4" and other improper fractions are simply "impossible numbers").

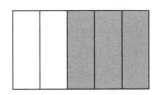

Fig. 2.3. Does this diagram show 3/5? Not necessarily.

Exploring the Mathematical Properties of Fractions

Once students can generate different quantities (continuous and discrete) for given fractions and fractions for different displays of quantity, they are ready to explore fractions as a system of numbers. Two challenging, important, and often fascinating aspects of fractions are their equivalence and order properties. In both cases, fractions defy students' intuitions from natural numbers: There are many different fractions (an infinite number) equal to any given fraction, and between any two fractions there are infinitely many others. Exploring these surprising properties and eventually mastering them will take many years. But the investment in time is worth it for two reasons: (1) Students will understand very little about the arithmetic of fractions if they do not master order and equivalence, and (2) the range of practical application of these ideas in everyday life is vast.

One way into this rich mathematical domain is through the simple question "Which of the following two (or more) fractions is greater, or are they equal?", using carefully selected pairs and groups of fractions. The phrasing of the question is important; it leaves it completely open whether fractions that "look different" really are or just seem to be. I have used this comparison task with much success to study students' understanding of fractions from fifth grade through the college years. I found substantial variation in how students solve the comparison task; there are many different ways of thinking about order and equivalence. Students with the strongest understandings can use many different approaches. To help students connect their constructed knowledge with what we hope to teach them, we should listen for these different approaches and strategies, support them when we hear and understand them, and help all students build a rich repertoire of fraction strategies.

Consider comparing "3/5" and "5/7." Depending on their age, students' first approach to this problem may be to think about the corresponding divided quantities or sketch them on paper. Because both denominators are prime numbers, this is a challenging task, especially if they start with a circular quantity. But upper elementary school students can partition square or rectangular wholes with care and effort. What makes this problem interesting is that imagining or drawing the quantities often does not always resolve the issue; for many "3/5" and "5/7" are just "too close together to tell." Some students, not satisfied with visual judgments from their diagrams, will also reason more analytically about the size and number of the parts in each. But with this approach, they may also reach an impasse: "There are three of the fifths and fifths are larger than sevenths, but there are more sevenths!" As with all good problems, resist resolving the tension that can build here. If some students think that "5/7" is greater because "it looks bigger," ask them to convince others. Someone will eventually see that both fractions (or diagrams) are "missing two parts to equal a whole." Since "fifths are larger than sevenths," "3/5" must be less than "5/7" because it is missing more. Also, someone may notice that it is useful to draw a line down the middle of each diagram to see "where one half is." This makes it easier to see that "3/5" is a bit more than "half" but "5/7" is still more (in effect, that "5/7 − 1/2 > 3/5 − 1/2" but in quantitative terms). This move may even help to show that "two and a half fifths" and "three and a half sevenths" are both "half."

From a series of carefully chosen comparison problems (and other tasks), I found that upper elementary, middle, and high school students think about fraction order and equivalence in four basic ways. Many students, especially the younger ones, need to construct and examine divided quantities. But in time they see that they can reason about the size and the number of parts they have constructed, despite the flaws in their diagrams. For example, they may get fed up with trying to draw out "elevenths" and "thirteenths" to decide if "7/11 or 7/13" is greater. In their frustration, they come to see that the only thing that matters is that "thirteenths" are smaller than "elevenths," so 7/11 > 7/13.

With lots of experience reasoning about fractions with divided quantity diagrams, students can shift almost seamlessly to reasoning directly about the given numerator and denominator. For example, with "7/11" and "7/13," some students soon conclude that 7/11 is greater because "11 is smaller than 13," and when asked, explain that "'thirteenths' are smaller than elevenths." What is going on here is a shift away from physical parts to the numbers that represent the parts. Though the concept remains the same, reasoning directly about the numerical component is more efficient. This shift from parts to components allows students to apply their extensive knowledge of whole numbers. When fractions are equivalent (e.g., "8/16" and "12/24"), students can use their knowledge of multiplicative relationships, like "twice" or "half of" and "three times" or "a third of." They will find and talk about these multiplicative relationships both "within" ("8/16 = 12/24" because "8 is half of 16" and "12 is half of 24") and "between" ("2/3 = 6/9" because "2 is one-third of 6" and "3 is one-third of 9"). This reasoning is usually grounded in their thinking about divided quantities. But should you become suspicious that a student is just reciting a rule, ask him or her to show why their reasoning works by drawing diagram.

The third way to think about fractions is where they lie relative to important numerical markers or "reference points." For proper fractions, these reference numbers are "0," "1/2," and "1" (though all whole numbers work in similar ways for fractions greater than 1). With a difficult comparison like "8/11" and "7/15," some students as early as the upper elementary school years will see that "8/11 is greater than 1/2 and 7/15 is less than 1/2," so 8/11 is greater. This noticing is often supported by the idea that "half" for any denominator can be found by dividing it by 2. If asked to explain, some would go on to say that "half of 11 is 5 and a half" and "half of 15 is 7 and a half," so "7/15 is not quite a half." Some students will extend this approach by thinking of 1/4 and 3/4 as reference points. The capacity to reason by proximity to numerical reference points also shows up in reasoning with large whole numbers as well. With fractions, it appears to grow out of students' extensive experience with divided quantities (e.g., "9 ninths is a full whole") and their early knowledge of "half" as a multiplicative relationship.

Finally, in the middle school years and beyond, students learn the numerical transformations for fractions, conversion to common denominator, conversion to decimals, and cross-multiplication as quick ways to decide questions of order. In contrast to the constructed strategies sampled above, these strategies are usually learned in school. The problem with these methods is that many students cling to them without understanding why they work. So a good classroom strategy is to make finding a way to explain why common denominator works an on-going problem. Easy "proofs" for conversion to common denominator are accessible to students through divided quantity and multiplication by the right form of 1, and cross-multiplication is simply a disguised part of that strategy. But other interesting conversions are possible: Try having students use a conversion to common numerator and have them explain to themselves why it works!

In presenting this overview of reasoning about fractions through (1) divided quantities, (2) numerical components, (3) reference points, and (4) numerical conversions, I have focused on the variety of strategies that students learn, both on their own and from our teaching. Students who learn to use many of these strategies are those with the strongest and more durable fraction knowledge—knowledge that will continue to support their growth for many years. But not all students' strategies are flawless.

Many of students' faulty strategies arise from the fundamental mistake of interpreting quotients as pairs of whole numbers (e. g., seeing "3/5" essentially as "3 and 5," in at least some contexts). Faulty "additive equivalence" strategies can appear right alongside "correct" multiplicative ones (e.g., "3/5 = 5/7" because $3 + 2 = 5$ and $5 + 2 = 7$) either within or between. More-complicated misconceptions emerge when our tasks are more demanding. For example, I have asked students if they think there are fractions between 3/5 and 5/7. Sarah, who had solved the same comparison (5/7 > 3/5) a few minutes earlier, identified many such "in-between" fractions: "3/6, 3/7, …, 3/12; 4/2, 4/3, 4/4, …, 4/12, and 5/2, 5/3, 5/4, 5/5, and 5/6." In trying to deal with differences in both numerator and denominator, Sarah appealed to whole-number order and organized fractions by numerator family, not denominator family. "12" was the largest denominator in her numerator families because rulers were divided in inches! All these faulty strategies have two elements in common. They show students trying to integrate their growing knowledge of fractions with what they know about whole numbers, and they appear right alongside correct strategies. Since many students who do not openly express these strategies are thinking in similar ways, these strategies should be considered—not reprimanded—in the classroom. Pointing students back to divided quantities and letting them apply their idea on accurate diagrams is the best general approach to correction.

BUILDING MEANING FOR FRACTION ADDITION

Even with the decreased emphasis on rational number computation in Standards-based curricula, the total number of hours devoted to teaching (and reteaching) fraction arithmetic in the middle grades is enormous. There are no easy recipes for teaching these difficult procedures, because challenging new ideas like the incommensurability of different units ("thirds" are the not same as "sevenths") underlie them. They take time and patience to master—for both students and teachers. But two general principles have proven useful: (1) vary the kinds of tasks posed to students and (2) build on the foundation of divided quantities—especially with addition.

Conceptually, the meaning of fraction addition follows from the meaning of adding (that is, combining) quantities. With two similar divided quantities, it makes perfect sense to consider their combined size. For positive rational numbers, addition is an increasing operation: Sums are greater than either addend. This property opens the way to estimation as a precursor to work with the standard procedure. Nearly every middle school teacher is familiar with "add across" error (e. g., "3/5 + 2/3 = 5/8"). Though students who make this mistake may be thinking about fractions as pairs of whole numbers (as above), the error is often justified by appeal to divided quantity: Three parts of 5 and two parts of 3 do, in some sense, combine to yield five parts of 8, but only by ignoring the relative sizes of fifths, thirds, and eighths. This misconception can be undermined directly by asking students to consider the relative sizes of the addends and the sum, but there are also excellent complementary tasks.

Instead of asking for the exact sum, ask instead for an estimate (e. g., *"About how much do we have if we add 3/5 and 2/3?"*). Since both addends are somewhat greater than 1/2 (reasoning by reference points), students can use their existing knowledge to infer that the sum must be greater than 1 and likely close to 1 1/2. The request for an estimate focuses students' attention more on what is being added (especially the size of the addends) than on the steps in a procedure. Asking for an estimate is also more likely to engage students' knowledge of the real-world quantities and their properties, which would also undermine "5/8" as a possible answer. They must learn to attend to the meaning of addition to correct their inevitable whole-number-based mistakes.

Explicit connections can be built between the addition procedure and the meaning of fractions by examining key steps in the addition procedure. Students naturally question why they have to change the denominators before adding and what makes that "OK." Of the two justifications accessible to middle school students, the first is based on the multiplicative identity: multiplying by a suitable form of "1" leaves the addends unchanged. It works for some students, but its connection to the meaning of fractions is obscure, and it leaves many with a sense that they have been tricked mathematically. The second approach connects the change in the addend denominators to the subdivision of parts in the addend quantities (as illustrated in figure 2.4). Students must still appreciate the need to add commensurate units (i. e., equal-sized parts), but now they can see how those parts are generated.

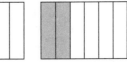

1/2 + 1/3 = ?
Parts are not the same size.

Multiplying 1/2 by 3/3 makes 1/2 = 3/6
Multiplying 1/3 by 2/2 makes 1/3 =2/6.
Now the parts are the same.

Fig. 2.4. Multiplication (raising terms) as the subdivision of parts

As students begin to work with and understand the addition procedure, consider reversing the direction of their thinking. Instead of giving them the usual series of addition problems (that is, pairs of addends), give them a series of sums and ask them to find the right addends from a set of possibilities. For example, "Find the addends for 2/3 from the following set, {1/2, 1/3, 1/4, and 1/6}. Instead of thinking algorithmically, they are more likely to think first in terms of the relative size of the addends (e.g., "2/3 is greater than 1/2, so maybe 1/2 is an addend and I need to look for something small to go with it"). Some students may tend to compute all the possible sums first, especially for small addend sets, but you can make that strategy more difficult by increasing the size of the addend set. This task can easily be extended to subtraction.

THE NATURE OF RATIOS

It is established practice in our mathematical culture to use quotients, numerals of the form *a/b*, to name ratios as well as fractions. Mathematics educators have struggled to agree on a shared meaning for ratio; indeed, no single consensus definition can be found in the mathematics education literature. I will use the term *ratio* to describe a relational number that has two properties: (1) it relates two quantities in one situation, and (2) it projects that relationship onto a second situation in which the relative amounts of the two quantities remain the same. For example, suppose a soccer team scores 11 goals in 5 games. Someone could think about the ratio "11 goals, 5 games" in order to estimate how many goals the team would score in a 15-game season (that is, 33 goals, 15 games). In the language given above, the first situation is "scoring in 5 games" and the second situation is "scoring by the end of the season." From this perspective, thinking of a ratio is equivalent to what many mathematics educators (and most textbooks) call *proportional reasoning*—with one important exception. People can think of a common ratio extending across the two situations and never write a fraction or an equation between two fractions. Proportional reasoning is a kind of thinking; it is not a matter of what the person writes down on paper.

Ratios are multiplicative, not additive, in nature. In essence, thinking with a ratio in the soccer-season context (that is, thinking of the two situations multiplicatively) involves replicating the 11 goals across each group of 5 games. If there are many such situations in which the quantities stand in the same multiplicative relationship, thinking about a ratio is equivalent to thinking about the linear function $F(x) = mx$ where m is the slope—a very important kind of ratio. In contrast, someone might think of the "goals to games" relationship in the first situation as "6 more (goals than games)."

Applying that to the second situation yields a predicted 21 goals scored (15 games + 6 more). This pattern of thinking (1) links the two situations (like ratio thinking) and (2) is comparative (like ratio thinking), but it is additive in nature, and a wide body of research shows how often middle grades students think additively and incorrectly about situations we see as proportional (e.g., Hart 1988; Lesh, Post, and Behr 1988). These students have conceptualized the same relationship between the two quantities in two different situations, but they have not conceived of that relationship as a ratio.

THE ORIGINS OF RATIOS

The central challenge of developing students' capacity to think with ratios (to reason proportionally) is to teach ideas and restrain the quick path to computation. The research literature on ratio and proportional reasoning shows that students' thinking in situations that we see as proportional often shows they are failing to think about "what is going on" in those situations. In their rush to compute an answer, students do not seem to think long enough about how the quantities in the situation relate and change. Fortunately, there is also evidence that students acquire important foundational ideas that can support more solid understandings of ratios later on.

It appears that some conceptual precursors of ratio appear quite early in children's thinking. Researchers have found that preschool children have intuitive notions of scale (the "large" object in one set belongs with the "large" object in another associated set) and of covariation (if one of two matched objects changes in size, the other must also). These ideas appear to support the later development of elementary school students' sensible additive strategies in proportional situations.

Lamon (1993) and Kaput and West (1994) have studied these additive approaches carefully in middle school students' solutions of "missing value" proportional problems. In both studies, students extensively used "building up" strategies to solve such problems before instruction in proportional reasoning and, to a lesser extent, after. "Building up" the quantities from one situation to equal the quantities in the second is additive thinking, because fundamentally students are adding quantities one group at a time. Although this is not ratio thinking, students are correctly reasoning about the situation—though not yet multiplicatively.

For example, when given the following problem (Kaput and West 1994, p. 245),

A restaurant sets tables by putting 7 pieces of silverware and 4 pieces of china on each placemat. If it used 35 pieces of silverware in its table settings last night, how many pieces of china did it use?

Sixth-grade students typically incremented the set of 4 pieces of china the same number of times as they incremented the set of silverware to reach 35 pieces (that is, 5 times to reach 20 pieces of china). Sometimes the strategy was executed by skip counting, sometimes by saying each ordered pair (e. g., "For 7 silver, there are 4 china; for 14 silver, there are 8 china;" and so on). But at the core, the strategy was a matter of building up larger quantities of silver and china from the combined unit, "7 silver and 4 china" (Lamon 1993). This research has also shown that students' capacity to handle such situations, either by additive or multiplicative reasoning, is strongly influenced by situational factors. Three main factors exert a strong influence on the relative difficulty of such problems: (1) the nature of the situation and students' experience with it (e.g., containment, price, and motion are easier than more abstract situations), (2) the sort of the numbers involved (small whole numbers are easier than larger numbers and rational numbers), and (3) the character of the ratios, within and between corresponding quantities (simple ratios like "2 times" or "3 times as large" are much easier than larger and non-unit ratios). Consequently, it makes little sense to talk about students' capacity to reason with ratios as a unitary ability. Young children, armed with their knowledge of "half," will solve familiar situations with nice numbers quite early; more-difficult situations will stymie students well into high school.

Lamon and Kaput & West also found that their students (mostly sixth and seventh graders) rarely used cross-multiplication to solve missing-value problems, even after the method was taught in class. Although cross-multiplication is both efficient and universally applicable (i.e., it does not depend on specific numbers in the problem), students either do not easily learn it or resist using it when they do. This may be due to the difficulty of linking it to their early knowledge of ratios. The procedure does not match the mental operations involved in the "building up" strategy, and more specifically, the cross-products (e.g., "4 china × 35 silverware") lack meaning in the situation. What, after all, is one china-silverware, much less 140 of them?

This research leaves us with one important question: How does early additive reasoning about proportional situations support the development of multiplicative reasoning, that is, thinking with ratios? On this crucial point there appears little consensus among researchers and educators, perhaps because the field of proportional situations is so vast. Though researchers are just now addressing this question, two sorts of cognitive capacities seem to support multiplicative interpretations of proportional situations: (1) the ability and willingness to partition and iterate composite units (e.g., if 4 pieces of china are matched to 10 pieces of silverware, seeing that 1 piece of china also matches to 2.5 pieces of silverware), and (2) a focus on relative, not absolute change in situations involving growth and accumulation.

Conclusion

Drawing on a diverse body of research, my central message has been that students bring a wide variety of experiences with fractions and ratios and knowledge borne of that experience to their work in our classrooms. Because of the nature of our physical and social world, much of this experience and knowledge centers on dividing (partitioning) and divided quantity. From this foundation grows much of the diversity of students' constructed knowledge and strategies. In our classroom teaching, we ignore this knowledge at our peril. Instead of treating our students as blank slates, we need first to find out how they think and what they know about fractions and ratios, and then begin the process of shaping, reshaping, and building on that knowledge. The cost of ignoring what they bring to our classrooms is that they will learn school procedures for solving school tasks and never integrate their constructed knowledge with the powerful new ways of thinking we have to teach them. If so, they will be poorly prepared to think mathematically in the increasingly complex world of quantities that awaits them.

References

Ball, Deborah L. "Magical Hopes: Manipulatives and the Reform of Math Education." *American Educator* 16 (Summer 1992): 14–18, 46–47.

Hart, Kathleen. "Ratio and Proportion." *Number Concepts and Operations in the Middle Grades,* edited by Jim Hiebert and Merlyn Behr, pp. 198–219. Hillsdale, N.J.: Lawrence Erlbaum Associates, 1988.

Jenkins, Lee, and Peggy McLean. Fraction Tiles. Hayward, Calif.: Activity Resources, 1972.

Kaput, James J., and Mary West. "Missing-Value Proportional Reasoning Problems: Factors Affecting Informal Reasoning Patterns." In *The Development of Multiplicative Reasoning in the Learning of Mathematics,* edited by Guershon Harel and Jere Confrey, pp. 235–87. Albany, N.Y.: State University of New York Press, 1994.

Lamon, Susan J. "Ratio and Proportion: Children's Cognitive and Metacognitive Processes." *Rational Numbers: An Integration of Research,* edited by Thomas P. Carpenter, Elizabeth Fennema, and Thomas A. Romberg, pp. 131–56. Hillsdale, N.J.: Lawrence Erlbaum Associates, 1993.

Lesh, Richard, Thomas Post, and Merlyn Behr. "Proportional Reasoning." *Number Concepts and Operations in the Middle Grades,* edited by Jim Hiebert and Merlyn Behr, pp. 93–118. Hillsdale, N.J.: Lawrence Erlbaum Associates, 1988.

Piaget, Jean, Barbel Inhelder, and Alina Szeminska. *The Child's Conception of Geometry.* New York: W. W. Norton, 1960.

Thompson, Patrick W. "Concrete Materials and Teaching for Mathematical Understanding." *Arithmetic Teacher* 41 (May 1994): 556–58.

3

Children's Development of Meaningful Fraction Algorithms: A Kid's Cookies and a Puppy's Pills

Janet M. Sharp

Joe Garofalo

Barbara Adams

MANY children have difficulty developing connected knowledge about fractions concerning number sense, operation sense, and algorithmic skills (Behr et al. 1992; Kieren 1988). Procedural knowledge, such as algorithms for operations, is often taught without context or concepts, implying to the learner that algorithms are an ungrounded code only mastered through memorization. Introducing algorithms before conceptual understanding is established, or without linking the algorithm to conceptual knowledge, creates a curriculum that tends to be perplexing for children to master or appreciate (Carpenter 1986). This untimely rush toward symbol manipulation fosters misconceptions about a lack of connectedness both between concepts and procedures and between fractions and students' everyday lives. However, once children have developed a conceptual knowledge base for fraction sense and operation sense, they can meaningfully learn, or even create for themselves, appropriate fraction algorithms.

Kieren (1988) hypothesized that children gradually expand their knowledge and thinking about fractions by building it up from personal environments. As children learn, they develop more intuitive knowledge in which they combine thought, informal language, and images. They become more able to extract mentally, and think about, fractional ideas without a strong dependence on the specific context. Eventually, they begin using formal

The authors wish to thank Leah and Stephanie for their help and cooperation.

symbols when they become able to connect concepts and procedures and to use and understand conventional language, notations, and algorithms.

TWO CHILDREN DEVELOP PERSONAL ALGORITHMS

We shall describe two girls who developed personal algorithms, based on conceptual knowledge for fractions. First, you will meet Joe and his daughter, Leah. Leah used sophisticated mental strategies to divide whole numbers by whole numbers and add fractions with different denominators. Second, you will meet Janet and her student, Stephanie. Stephanie used explicit pictorial and symbolic strategies to divide whole numbers by fractions. Both girls developed their strategies through encounters with real-world division situations and with encouragement from adults.

Kids and Cookies

When Leah was almost five years old, Leah and Joe often played games while driving to Leah's preschool in their pickup truck. In one game, which Leah named "Kids and Cookies," Joe posed situations and questions to help Leah establish conceptual understanding of division and fractions. He would give her a number of kids and a number of cookies, and ask Leah how she would share the cookies. At first Joe gave Leah specific, real-to-Leah situations that led to whole-number responses (i.e. What if Molly and Lee were at our house and we had 6 cookies, how would you share the cookies?), and then later he presented her with less descriptive and less personal situations that led to fractional results. Below is an excerpt of a conversation that took place one morning (reconstructed after Joe arrived at his office):

Joe: Hey Leah, what do you want to play today?

Leah: Let's play Kids and Cookies.

Joe: OK. What if you had 4 cookies and 3 kids? How would you share them?

Leah: One, one, one, and then there is one left. Then they each get one third, one third, one third.

Joe: So, how much does each kid get?

Leah: They get one whole one and one third.

Joe: What if you had 5 cookies and 3 kids? How could you share the cookies?

Leah: One, one, one. Then there's two more left. OK. Then, they get a third, a third, a third, and then a third, a third, a third.

Joe: So, how much do they each get?

Leah: They get one whole and two thirds.

Joe: What if you had 7 cookies and 4 kids?

Leah: That's a hard one, maybe I can't do it.

Joe: Do you want to try?

Leah: Yeah, but it's hard.

Joe: Think about what you did to solve the other two.

Leah: Whole, whole, whole, whole, then there's three more left. Um, three more cookies left. Then you break up one into halves, then there are two left. And, another into half, half. Break the last one into quarter, quarter, quarter, quarter.

Joe: Great! How much does each kid get?

Leah: One whole, one half, and one quarter.

Joe and Leah played the game for a few minutes, several times a week, in their pickup for several months, with Joe varying the numbers so that Leah was able to resolve more and more complex situations. Once Leah developed conceptual understanding of fractions in this context, and her own method of resolving situations, Joe introduced her to fraction notation. Figure 3.1 shows Leah's work (age 5) on one of the situations she previously worked out in the truck. Note that under each kid's face, she has listed the fractional amounts each child received—1/4, 1/2, 1.

THERE ARE 4 KIDS AT A PARTY AND THERE
ARE 7 BROWNIES FOR THEM TO SHARE. HOW
CAN THEY SHARE THEM?

Fig. 3.1. Leah's work

Often Leah would be asked to determine two ways to share the cookies (e.g., 4 cookies could be shared among 6 kids by splitting each cookie into sixths and giving each kid 4 sixths, or by giving each one half of a cookie and one sixth). By comparing the two ways, Leah developed an understanding of equivalent fractions.

Daddy's Pies

Leah and Joe made up a subsequent game, which Leah (age 6) named "Daddy's Pies." Joe gave Leah fractions of several pies he ate and asked her to figure out the total amount of pie he had eaten. After Leah resolved a few situations involving common denominators (e.g., 1/4 of an apple pie and 1/4 of a cherry pie) and a few situations involving halves and quarters (e.g., 1/2 of an apple pie and 1/4 of a cherry pie), Joe posed the following:

> *Joe:* If I ate one half of an apple pie and one third of a cherry pie, how much did I eat?

Leah at first struggled with this task, but after being asked to compare it to those she previously solved, thought for a while, and replied, "I have to break them into same-size pieces." She briefly considered fourths, but then tried sixths. She was able to "break" 1 half into 3 sixths and 1 third into 2 sixths, and eventually responded, "5 sixths." Later that day, one of Joe's friends asked Leah to explain how she solved the problem. After she articulated her method, he asked:

> *Frank:* What if your daddy ate one half of an apple pie and two thirds of a cherry pie?

Leah thought about it for a moment and then left the room. She returned a few minutes later and proudly announced "seven sixths." After resolving one or two similar situations, a few times a week for the next two weeks, Joe posed the following:

> *Joe:* If I ate one half of an apple pie, one third of a cherry pie, and one quarter of a peach pie, how much did I eat altogether?

Leah was able to generalize her method to resolve this situation. She first determined that the "same-size piece" was a twelfth, and then broke each amount of pie into twelfths by determining how many times each denominator went into twelve. For one half, she counted out loud, while keeping track of her counting on her fingers, "2" (extend 1 finger), "4" (extend a second finger), "6" (a third finger), "8" (a fourth finger), "10" (a fifth finger), "12" (a sixth finger) "makes 6 twelfths." She then had 6 fingers extended, corresponding to the six numbers she had counted out, which was the numerator on her equivalent fraction, 1/2 = 6/12. Similarly for one third, she counted out, again with her fingers, "3, 6, 9, 12, makes 4 twelfths." For one fourth, she counted out, "and 4, 8, 12, makes 3 twelfths." Then she added "Six twelfths plus 4 twelfths is 10 twelfths ... plus 3 twelfths is 13 twelfths. So you ate one and one twelfth."

She was later able to extend this method to situations involving fractional amounts such as five sixths of a peach pie by determining how many times a denominator went into the common denominator, and then multiplying the numerator by that quotient. For example, for her dealings with five sixths, she counted out, "6, 12... that makes 2 twelfths, and 2 times 5 gives 10 twelfths." After she had mastered her method of "breaking into same-size pieces," Joe introduced her to the term "common denominator" and explained how she would be expected to write solutions on paper in school.

A Puppy's Pills
Quotients without remainders.

Now meet nine-year-old Stephanie. By fourth grade, Stephanie had learned to find equivalent fractions and to recognize fractional situations. Janet worked with Stephanie's classroom teacher, Barbara, for a three-week period during their study of division of fractions. Since Janet was working in a class of twenty-four students, she posed a situation to the class that was personal to her, rather than attempt to find a situation that could be personal to each of the twenty-four children. She posed the problem in figure 3.2.

The children were devastated. They wanted reassurance that the puppy had recovered. Questions like, "What is the puppy's name?", "Is she okay now?", and "Did the doctor say what was wrong?" demonstrate the children's empathy and subsequent personal investment in the situation. Only after they were satisfied as to the health of the puppy, did the children begin solving the problem.

Stephanie resolved this situation by drawing 15 circles broken into thirds. Then, she subtracted, encircling groups of 5/3 on her drawing, until all of the pills were gone (9 days). Stephanie recorded the problem as 15 ÷ 1 2/3 = 9. From her picture, Stephanie seemed to understand some things about fraction division. She recorded the splitting of the pills into thirds (common denominator) so she could repeatedly subtract (divide) and count her regroupings. She had completed the task, found the correct answer, and used a nice strategy. So, Janet guided Stephanie to work with formal symbols by suggesting, "Let's record our work so someone who did not have the pictures could follow our thinking." Stephanie recognized, explained, and recorded all her thinking.

When I got home last night, I found my puppy not feeling very well. So, I took her to the veterinarian

Our vet said to give our dog some medicine. She gave us 15 tablets.

Because our dog is very large (100 pounds), the vet said to give the dog $1\frac{2}{3}$ tablet each day. For how many days will the medicine last?

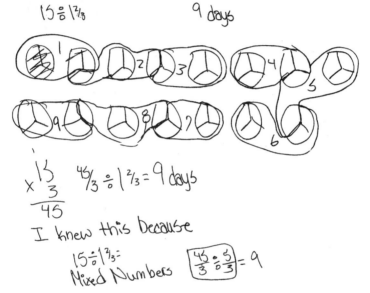

Fig. 3.2. Stephanie's work

Janet: What did you do after you drew the 15 pills?

Stephanie: I split them into thirds.

Janet: I wonder how you could record that with numbers.

Using her picture, Stephanie thought for a while, multiplied (15×3), and wrote: 45/3.

Janet: Now, I want you to think about how much you subtracted for each day.

Stephanie: I subtracted five of those thirds until they were all gone.

Janet: How could you write that?

Stephanie then wrote: ÷ 5/3. (She knew that division was repeated subtraction.)

Janet: I wonder how you solve *that* problem?

Stephanie looked at her work: 45/3 ÷ 5/3 and her solution of 9 days. Then she verbally made an observation, "Isn't 9 just 45 ÷ 5?"

Quotients with remainders.

Next, the class was given a variety of division of fractions problems that did *not* result in whole-number solutions. These complex problems can encourage the development of procedures for resolving situations.

For example, the problem "What if our dose had been 1 3/4 tablet each day?" resulted in a "remainder" and caused much discussion among the children, including Stephanie. Stephanie saw that she could give the puppy a complete dose of medicine for 8 days. But, then a whole pill remained. Like many of her classmates, Stephanie had drawn pictures of circles split into fourths (common denominator), and then repeatedly subtracted (division) by drawing rings like she had done in the previous problem, until she had only the 1 whole pill left. At first, Stephanie was satisfied to resolve the situation by suggesting that there was a remainder of 1, which is accurate. Several students suggested throwing the pill away. Others suggested giving the pill to the puppy since it was less than the dosage and should not hurt the puppy. Each of these is, of course, a possible real-world solution to the situation!

But in the school-world, we pushed for a mathematical solution to the situation. Stephanie was now at a crucial stage, if she was to generalize an algorithm for division of fractions. She could not be allowed to whisk the idea of the remainder into the background, even though throwing out the excess pill was a real-world solution. To generalize, Stephanie needed to be encouraged to think in terms of fractional portions. Janet revised her question, "What portion of a day's medicine dose remained?"

Stephanie looked at her picture and said, "Well, it takes 7 parts [fourths] to make 1 medicine dose, and here I have 4 parts left. So, it must be 4/7 of a dose." She wrote: 15 ÷ 1 3/4.

Then, she used a familiar whole-number tool to do the procedure:

$$
\begin{array}{r}
8 \\
7\overline{)60} \\
56 \\
\hline
4
\end{array}
$$

And, she recorded the following fractional symbols for her thinking: 60/4 ÷ 7/4 = 8 4/7.

DISCUSSION

Conceptual Knowledge Base

For Leah, the concept of a fraction grew from encounters with whole-number division problems. She handled remainders by sharing portions of cookies and pies. These contexts were real-world, highly specific to Leah's personal life. In the cookie situation, the followup questions Joe asked incorporated ideas about sharing and used Leah's informal language. These meaningful questions together with the lack of paper and pencil encouraged her to concentrate on her own images. Also, Leah connected ideas about the addition of fractions to the need for a common denominator when she noted that it would be easier to add pieces of pie if they were broken into same-size pieces! Because they were driving in a truck, Leah was forced to develop sophisticated mental strategies to keep track of these common denominators.

Stephanie used paper and pencil to draw pictures of the pills and then she manipulated the drawings to complete the subtraction process. She already knew about concepts related to equivalent fractions and pictorial representations, as well as associated symbolic representations. Her conceptual knowledge about fractions also enabled her to recognize the need for a common denominator to subtract fractions. Using the pictures allowed her to determine what the "same-sized pieces" would be without using symbol manipulation. Then she completed her fractional division problem, much like she would have completed a whole-number division problem, by dividing the numerators.

Developing Personal Procedures

Clearly, both Leah and Stephanie used their personal fraction sense to resolve these situations. Leah followed the "fair-sharing" definition whereas Stephanie used "repeated subtraction." However, in both cases, the girls encountered a division-with-remainder context. This approach grounded both girls' thinking in a familiar conceptual base that used a very real-world situation and that built on their existing knowledge.

To resolve Kid's and Cookies situations Leah developed a meaningful procedure to share cookies that was based on her understanding of fractions. This procedure was first to distribute evenly as many whole cookies as possible, then to cut remaining cookies into smaller fractions as needed and distribute them evenly, and finally to list the pieces given to each kid. To add the pieces of Daddy's Pies, Leah created her own way of breaking fractions of pies into same-size pieces. This procedure involved calculating the numerators of "same-size piece" fractions by first determining how many times the denominators of the original fraction divided into the common denominator.

In order to resolve the Puppy's Pills situation, Stephanie used her conceptual understanding of division with whole numbers (as repeated-subtraction) to develop knowledge about how to divide fractions. The followup question that resulted in a remainder was designed to promote the development of a procedure. Indeed, Stephanie used a known whole-number division process to solve it. Throughout this effort, she manipulated her mental images to build and refine her knowledge, as Kieren (1988) suggests should happen. However, she did need prompting to record her thinking consistently with fractional symbols.

Recommended Teaching Practices

1. Use Personal Contexts to Develop Concepts.

Children engage in meaningful thinking about fractions when the contexts relate to their personal situations. (Such contexts motivated Leah and Stephanie.) Carefully selected contexts lend themselves to followup questions the teacher needs to ask. In appropriately guiding children, the teacher should prepare the children to build on conceptual knowledge in their own unique ways. Conceptual knowledge then provides a base on which to develop procedural knowledge. So, it is important to select situations that tend to promote procedure development.

2. Encourage Invented Procedures.

Once some conceptual knowledge has been established, children must be encouraged to use that knowledge to travel their own paths and develop meaningful, personal algorithms (Kamii and Warrington 1995). Teachers should be cautious neither to rush the child's invention of an algorithm nor to push the child toward a preconceived algorithm. Children often make mistakes as they work to develop a procedure. The temptation to insulate children from making a mistake, by showing them how to do an operation, is enticing because we are then romanced into the illusion that students who can reproduce our action have "learned" it. The teacher's job is patiently to monitor and direct the child's invention of a useful algorithm. The teacher must take care to ensure that procedures eventually created by the student are correct procedures and that the child has not overgeneralized some procedure inaccurately.

3. Allow Them to Use Their Own Language and Pictures.

Teachers should facilitate discussion and thinking about the problem using the children's informal language and work to develop students' abilities to form mental images. Some students may not immediately make generalizations with numerals. However, they can make early generalizations with images and pictures. Children's uses of pictures enable them to understand and resolve situations and perform procedures they might otherwise find beyond their grasps. Pictures also allow crucial issues, such as the importance of common denominators, to become clear at a conceptual level.

4. Encourage Children to Keep Track of and Record Work.

When dealing with their invented procedures, language, and pictures, children need to keep track of their procedures (Garofalo et al. 1989), whether that track-keeping is kinesthetic, such as Leah's extended fingers, or with paper and pencil, such as Stephanie's drawn circles. In order to generalize the procedures, the children must eventually be encouraged to record their thinking with paper and pencil. The written work may initially be informal, but can be used as a springboard to the more formal symbolic representations of procedures.

5. Move to Formal Language and Algorithms.

Once children have developed their ideas in personal, informal ways with mathematical integrity, teachers should help students connect their ideas and procedures to conventional algorithms. Classroom discourse should be rich with conventional symbols and language. Lessons should connect concepts and procedures and culminate in the use of symbols.

FINAL COMMENT

Leah and Stephanie are typical students who have had atypical learning experiences, which gave them opportunities to showcase unexpected thinking. Whether learning 1-on-1 or 24-on-1, careful use of personal contexts, building procedures on concepts, and patient listening will encourage children to respond with good thinking. Both girls encountered problems for which they had no previously learned algorithms. Each built on her existing conceptual understanding of fractions and operation sense of whole numbers to develop procedures to resolve real-world situations involving fractions. Two of the situations required the girls to make sense of addition or subtraction with a least common denominator, and each did this in her own way at first. Eventually, they both developed procedures from their existing conceptual knowledge.

REFERENCES

Behr, Merlyn J., Guershon Harel, Thomas Post, and Richard Lesh. "Rational Number, Ratio, and Proportion." In *Handbook of Research on Mathematics Teaching and Learning*, edited by Douglas Grouws, pp. 296–333. New York: Macmillan, 1992.

Carpenter, Thomas P. "Conceptual Knowledge as a Foundation for Procedural Knowledge." In *Conceptual and Procedural Knowledge: The Case of Mathematics*, edited by James Hiebert, pp. 113–32. Hillsdale, N.J.: Lawrence Erlbaum Associates, 1986.

Garofalo, Joe, Vanere Goodwin, David Mtetwa, and Bob Mitchell. "Some Observations of Seventh Graders Solving Problems." *Arithmetic Teacher* 37 (October 1989): 20–21.

Kamii, Constance, and Mary Ann Warrington. "Division with Fractions: A Piagetian, Constructivist Approach." *Hiroshima Journal of Mathematics Education* 3 (March 1995): 53–62.

Kieren, Thomas E. "Personal Knowledge of Rational Numbers: Its Intuitive and Formal Development." In *Number Concepts and Operations in the Middle Grades*, edited by James Hiebert and Merlyn J. Behr, pp. 162–81. Hillsdale, N.J.: Lawrence Erlbaum Associates, 1988.

ADDITIONAL READING

Lappan, Glenda, and Mary K. Bouck. "Developing Algorithms for Adding and Subtracting Fractions." In T*he Teaching and Learning of Algorithms in School Mathematics*, 1998 Yearbook of the National Council of Teachers of Mathematics (NCTM), edited by Lorna J. Morrow and Margaret J. Kenney, pp. 183–97. Reston, Va.: NCTM, 1998.

Pothier, Yvonne, and Daiyo Sawada. "Partitioning: The Emergence of Rational Number Ideas in Young Children." *Journal for Research in Mathematics Education* 14 (July 1983): 307–17.

4

Organizing Diversity in Early Fraction Thinking

Susan B. Empson

IMAGINE you are teaching third grade, and you begin a unit on fractions by giving your students this equal sharing problem: *4 children are sharing 10 brownies so that everyone gets the same amount. How much brownie can 1 person have?* You ask your students to solve the problem using a strategy that makes sense to them. There are some tools available for those who want to use them: paper rectangles that can be cut or folded, linking cubes, and paper and pencils for drawing. Because your students have already solved multiplication and division story problems with whole-number answers, they are confident as they start to work on this problem.

Soon you see several things happening. One child, Sanchez, is using linking cubes to solve the problem. He deals the cubes out until there are two left, then he shows with a gesture of his hand how he would cut each of those cubes into four pieces. "Then everyone would get two more halves," he says. You note his use of "halves" for fourths. Another child, Gabrielle, is also using linking cubes to figure that each sharer gets two. Then she draws the remaining two brownies, clustering four people around each brownie (fig. 4.1.a). "I would put the people with the brownies," she says. "Everyone gets two more pieces." A third child, Obie, has written a division problem (fig. 4.1.b), with an answer of "2 r 2." A fourth child, Marina, simply tells you, "they get three altogether." As you continue to walk around, wondering how on earth to deal with the myriad issues represented by these responses, you see Kenny's answer and his drawing: "2 in a haf" (fig. 4.1.c, sic). And finally, you hear Arly exclaim: "You can't do it! It doesn't come out even." Nobody has used the paper rectangles that you cut out the night before. You suggest to Arly that perhaps she could use them to solve the problem.

The first-, second-, and third-grade classrooms in which I have worked, have had all these things happen, often in the same lesson. Although a fundamental tenet of the reform movement in mathematics education is to pose

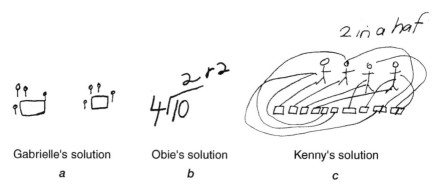

Gabrielle's solution   Obie's solution   Kenny's solution

a   b   c

Fig. 4.1. Children's solutions for 10 brownies shared by 4 children

problems for which students produce a variety of strategies or answers (e.g., Hiebert et al. 1997), sometimes that diversity can be daunting. But diversity is also the engine of change. The dilemma for you as a teacher is how to respond to the range of things your students do to solve problems and express their thinking, and especially, how to find the mathematics in it. What aspects of students' activity should you pay attention to? What aspects should you ignore? When should you pay attention to what? What should you show students, and what should you let them invent for themselves? What role can your questioning of children play in organizing this diversity of thinking? These questions have come up repeatedly in my work with teachers.

I do not believe that a teacher can or should address, in one lesson, all the potential mathematical issues that these third graders' strategies and explanations suggest. Teachers need a way to interpret students' thinking and to make principled decisions about what aspects to carry forward, and when to do it. In this article, I suggest a framework for making sense of students' activity as it relates to solving and discussing equal-sharing problems. The framework is based on the idea that students' thinking about fractions develops along some key dimensions, and that the coordination of these dimensions in class discussions can lead to increased mathematics understanding. I think of these dimensions as different aspects of problem-solving activity that become, through classroom interactions, interconnected and mutually enriching.

Children's *strategies* are the foundation and starting point. They are what children do to solve problems. *Representational tools* are used in close concert with strategies. They are the means to solve problems and express thinking. *Language patterns* refers to how children talk about what they have done, as well as the questions teachers ask and why they ask them. Together these dimensions of classroom interactions constitute the means by which the diversity of thinking expressed by students is organized into a meaningful body of mathematical knowledge.

The Development of Strategies for Equal Sharing

Children as young as first graders have informal knowledge of cutting things up that provides the basis for invented strategies for equal-sharing problems (Empson 1999; Hunting 1991). Strategy is the primary dimension of development because student-generated strategies can (and I believe should) serve as the foundation for mathematics instruction. A focus on student-generated strategies allows a teacher to begin with, and build on, what children already know, and it allows children to participate in instruction by making contributions that are personally meaningful.

Setting aside issues of language and representation, virtually all the children in our imaginary third-grade classroom have viable strategies for solving the equal-sharing problem. Further questioning reveals that they know, or are close to knowing, what to do to share the extra brownies. When you ask Marina, "How did you figure out that each person gets three?" she responds, "I gave them each two whole brownies, then I had to split the last two in half, so everyone could get one more." When you ask Obie what "2 r 2" means, he replies "I know it's division because it's splitting up brownies. There's two left over that I can't give out because it wouldn't be fair." You ask, "Is there anything you can do to share the last two brownies so that it would be fair?" He pauses, thinking, then says, "Well, I guess I could cut them in half."

Children have robust knowledge of halving and repeated halving, and so the equal-sharing problem that you have given your third graders is one that many young children can solve before any instruction in fractions (Empson 1999). Equal-sharing problems involving two, four and even eight sharers are good tasks to begin fraction instruction, because students will be able to generate strategies for such problems using their informal knowledge of repeated halving.

As the number of sharers in an equal-sharing problem changes from those that lend themselves to repeated halving (e.g., two, four, eight), to other numbers such as three or six, there is more variation in the strategies children use. (Note that it is not only the number of sharers that matters, but its relationship to the number of items shared.) For example, after your third graders have explored problems that can be solved by repeated halving, you might give this problem: *6 children are sharing 4 candy bars so that each one gets the same amount. How much can each child have?* Some children will continue to use repeated halving. Others will relate the number of partitions they make to the number of sharers; coordinating the partitions with the number of sharers represents more-advanced thinking.

Children's strategies for equal sharing develop according to how the child coordinates the number of shared items with the number of sharers to solve the problem (table 4.1). Coordinating the partitions of shared items with the number of sharers is important because it results in an exhaustive partition. That means everything in the sharing situation is completely shared, with no leftovers. Sometimes children will begin a strategy thinking about how to coordinate partitions with sharers, and other times it seems to happen almost as an afterthought, as they are nearing the end of a partition (e.g., fig. 4.2.). Eventually children conceptualize a direct connection between fractional quantities and ratios in equal sharing situations (Streefland 1991). For example, two brownies for every three children means everyone gets two thirds of a brownie.

TABLE 4.1
Development of Children's Equal-Sharing Strategies

Early Strategies	Description	Example
a) Repeated halving	Child repeatedly halves each unit, regardless of number of sharers. Little or no coordination with number of sharers	*3 children sharing 2 candy bars*
b) Trial and error	Child tries various partitions with little or no coordination with the number of sharers. Some children may go through a list of fractions (e.g., halves, thirds, fourths) until they find one that yields the right number of pieces to deal out.	*10 children sharing 3 candy bars*

Intermediate Strategies	Description	Example
c) Give out halves	Child starts by giving out halves, if possible. The rest of the partition is coordinated with the sharers in some way.	*6 children sharing 4 candy bars*

Table 4.1 (cont.)

Intermediate Strategies	Description	Example
d) Coordinating sharers with single units	Child partitions each shared unit into enough pieces for all sharers. (This is a useful, all-purpose strategy, within the zone of understanding of many first and second graders.)	*6 children sharing 4 candy bars* 1 sixth + 1 sixth + 1 sixth + 1 sixth
e) Coordinating sharers with multiple units		
1) coordinates total sharers with every two units.	Child partitions every 2 units into enough pieces for all sharers. There may be a leftover unit to partition.	*12 children sharing 9 candy bars* "I know that 2×6 is 12, so I can cut 2 candy bars in sixths, and keep doing that. There is 1 candy bar left over and can cut that in twelfths."
2) coordinates total sharers with every 3, 4, 5 or more units.	Child partitions every 3, 4, 5, or more units into enough pieces for all sharers. There may be leftover units to partition.	*12 children sharing 9 candy bars* "I know that 4 3s is 12, so I'll cut 4 candy bars in thirds and see what happens." 1 third + 1 third + 1 twelfth

Later Strategies	Description	Example
f) Coordinating sharers with all units		
1) creates same number of pieces as sharers.	Sometimes children try to create partitions that give each sharer exactly one piece. This means they have to use multiplication, division, or trial-and-error skip	*24 children sharing 8 candy bars.* "8 times what equals 24? I'll cut each candy bar in thirds."

Table 4.1 (cont.)

Later Strategies	Description	Example
	counting to figure out how many pieces to partition each unit into. Some children think of this as creating big pieces. This strategy is related to the idea of reducing to a unit fraction. It does not work in all equal-sharing situations.	
2) creates a number of pieces that is a multiple of the number of sharers.	This sophisticated strategy is used mainly by children who are fluent with multiplication. The child's goal is to create a number of pieces greater than the number of sharers that can be equally distributed among the sharers.	*4 children sharing 7 candy bars* "I want a number that both 4 and 7 can go into. If I cut each candy bar into 4 pieces, I'll have 28 pieces altogether, and each child can have 7 pieces or 7 fourths."

Sometimes it is not clear how children are thinking about the number of sharers as they decide how to cut things up. You can find out by asking questions like, "Why did you split these brownies into four pieces?" One child might say, "Well, I just kept cutting, and it worked out." Another might say, "Because there were four children." As children continue to solve and discuss equal sharing problems, you will see them using knowledge of doubling and halving, skip counting, multiplication fact families, and

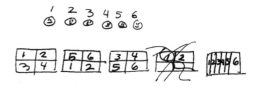

Fig. 4.2. Example of a repeated-halving approach to sharing 4 brownies among 6 children

ratio concepts (Empson 2001). This knowledge is revealed in the way children use it to coordinate the number of people sharing with the units (e.g., brownies) to be shared, and in the ways they use representational tools.

Children's Use of Representational Tools

A second dimension in the framework is *representational tools*. I include in this category children's drawings and diagrams, symbols, and manipulative-type tools such as paper rectangles, linking cubes, and pattern blocks. These are the things that support children's strategies and help make those strategies visible to others. Representational tools should be a combination of student-generated and teacher-provided. On the one hand, teachers may need to tell children about some of the conventional aspects of representation, such as fraction symbols. On the other hand, the same principles that support the argument for student-generated strategies apply to student-generated representations of fractions. Because strategies are closely coupled with the use of representations, encouraging the development of student-generated representations can enhance and deepen the meanings of representations.

When it comes to fractions, it is not unusual for textbooks to emphasize part-whole representations and fraction symbols, to the exclusion of other forms of expression. Allowing children to choose their own tools and make their own representations to solve equal-sharing problems, however, can foster an interesting diversity of thinking, which can contribute to richer understanding of the mathematics of fractions.

In our hypothetical third-grade classroom, for example, children are in the habit of using the tools of their choice to solve problems (e.g., Hiebert et al. 1997, chapter 7). Sanchez used linking cubes to distribute the brownies to sharers. He described how he would cut the last two brownies in fourths, and he showed where he would make the cuts using his hand as a knife. He does not show the partition using a part-whole representation. Instead the emphasis for him is on the action of cutting, and the fact that each sharer gets two whole brownies and two pieces. Gabrielle deals people to brownies. Like Sanchez, she does not represent how the brownies should be partitioned. Rather it is implicit in her repeated distribution of people to brownies. Unlike Sanchez and Gabrielle, Kenny uses a part-whole model to represent the partitioned brownies, and he shows how he would cut them by drawing the lines of partition. His pieces are uneven, however. When you point out that some pieces are bigger than others, he says, "I know. But I want the pieces to be all the same, because then it's fair." Finally, Arly has readily folded two paper rectangles to make perfect half brownies. She was meticulous about lining up the folded-over edges to make equally sized pieces.

Each of these uses of representational tools provides opportunities to discuss different mathematical aspects of fractions. For example, Sanchez knows that each sharer gets more than two brownies, but less than three, because of the way he distributes the cubes. Gabrielle's dealing-people approach highlights a ratio interpretation of fractions. She put four people for every one brownie, and repeated that until the people had shared all the extra brownies. The conception of fourths embodied in this approach is that when four people share one thing, they each get one fourth.

Many textbooks illustrate fractions using geometric shapes partitioned into equal parts. As intuitive as these part-whole representations seem, children do not produce them spontaneously when asked to draw to solve equal-sharing problems. Nonetheless, they can be productive, and they help children begin to think about some of the geometric aspects of partitioning fractions (Pothier and Sawada 1983).

More important, as children draw part-whole representations of other fractions, such as thirds, fourths, and sixths, attention to exactly how they make the partitions can provide a basis for a discussion of equivalence. For example, children often draw fourths by partitioning a shape in half, then in half again. A fourth can be described in the context of this process as "a half of a half." Sixths can be drawn in three ways: by making halves, then partitioning everything in thirds; by making thirds, then partitioning everything in half; and by making a guess about how big a sixth is and partitioning the pieces one by one. The first way can support children's justification for why one half and three sixths are the same amount; the second way supports a justification for the equivalence of one third and two sixths; and the third (fairly common) lends itself best to a consideration of the fact that six sixths is the same as one. Just as it is important to ask your children why they decided to make a certain number of partitions, it is also important to ask them how they made the partitions, and to use that information to lay a foundation for developing equivalence ideas.

Language Patterns

The way teachers talk with students about the fractional amounts they have created is central to the development of children's fraction conceptions. In contrast to strategies and representations, children do not readily generate fraction terminology. Students' tendency is to call all fractional pieces "halves" or "pieces." As you teach your unit on fractions, you may encounter several children who persistently call any fraction piece a half. Other children, like Marina, simply count fractional pieces as wholes, even though they recognize differences in size.

Your questions represent an opportunity to orient students to certain ways of thinking about fractions. As children begin their study of fractions through problem solving, the most salient aspect of equal sharing to them is the action of partitioning, not the sizes of the resulting pieces (nor for that matter the relationship between number of sharers and number of shared units). Students talk about this process in terms of their partitioning actions: "I split this in threes, and each person got another piece (or half)."

When teachers question students about fractional pieces, they often ask, "What is that piece called?" This kind of question frames fractional quantities as something distinct from whole-number quantities. We would not ask a child about a group of five cookies, "What is that called?" Similarly, the phrasing "one out of three equal pieces" supports a conception of fractions as two-part numbers.

Other lines of questioning that draw attention to the quantitative aspects of fraction can be more mathematically productive. For example, asking a child "How big is that piece?" or "How much brownie does one person get?" emphasizes fractional pieces as amounts of "stuff." A good question for helping children to articulate and conceptualize the relation between a piece and its whole is "How many of that piece would fit into the whole?"

Gabrielle's strategy raises the possibility of introducing slightly different language. Because she clustered four people around each brownie, you can quite naturally talk about "one brownie for every four people," chunking brownie and people into a ratio unit.

The early major focus in primary-grades fractions instruction should be on children's reasoning about equal sharing, rather than on using correct terminology. It is not necessary to introduce formal fraction terminology immediately in your unit. Some children will want to begin using it right away, and it can be casually introduced to them; many others will be more comfortable simply creating representations (e.g., folded or cut rectangles of paper, pictures of cupcakes that have been partitioned) and describing their actions. Their strategies do not depend on knowing correct terminology. Those children can indicate the size of a share by shading in or pointing to the appropriate amount.

ORGANIZING CHILDREN'S THINKING IN DISCUSSION

In our hypothetical third-grade classroom, it is time for children to share and discuss their strategies for 10 brownies shared by 4 people. This part of the lesson is crucial, for it is here that you will decide what aspects of children's thinking to bring forward, relate, and extend. It is also in this setting that students will have an opportunity to reflect on their own thinking and consider it in light of other students' thinking.

First you ask Sanchez to share. The other students cluster around his table. He demonstrates his strategy with linking cubes, and he explains how he would cut the last two brownies in four pieces each. "So," he concludes, "they each get two whole brownies, and two halves of a brownie." As he finishes his explanation, you ask him, "Why did you cut each brownie in four pieces?" "Because there were four people," he replies. Before you can ask the class whether they agree or disagree with Sanchez's way of solving the problem, Kenny interjects, "I got halfs, too, except I cut them different." You recognize an opportunity to discuss how the children's use of terminology relates to the fractional pieces they have created. Kenny presents his strategy on the board and you ask, "Could you color in how much one person is going to get?" You ask the rest of the group whether anyone else cut their brownies in four pieces, like Sanchez. Gabrielle and a couple of other children raise their hands. You glance at their papers, looking for a part-whole drawing of fourths to share. You find one, and ask the child to present her strategy next to Kenny's (fig. 4.3). There is some discussion about how her strategy is like and not like Sanchez's strategy with cubes. You leave both drawings on the chalkboard, because you want the children to be able to refer to them later.

Now that you have fourths and halves up on the board, you have two goals. One is to compare a one-half piece to a one-fourth piece and find out what the children say about the difference between them. Their descriptions will constitute the beginning of differentiating between pieces on the basis of their size relative to the whole. The second is to push your students to think about whether one half and two fourths are the same amounts. Because you have the two part-whole representations on the board, you will be able to refer to the pictures rather than using the fraction terminology. You will be listening for children to go beyond a visual comparison and talk about the relationship between halves and fourths. You have noted Sanchez's use of "half" to refer to a fourth, and you will listen in particular to what he contributes or seems to be taking away from the conversation. You now turn to Kenny and ask, "How are your halves different from Sanchez's halves?" and the lesson continues.

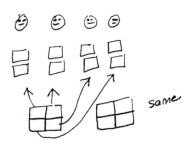

Fig. 4.3. A solution for 10 brownies shared by 4 children that uses a part-whole representation of fourths

One purpose of your questions in these exchanges is to probe students' thinking and to help them better articulate it. Sometimes you revoice what they have said, providing language that you believe will prove a useful tool for articulating their thinking later. You do not spend a lot of time on the details that children have gotten wrong, because there is so much they have gotten right, and you want to emphasize that point. You plan to address those issues later, as the need for more precise language and representation arises, and children's strategies are solidly developed.

Another purpose of your questions is to help children make connections between their strategy and other children's strategies, to extend their understanding. You will continue to write equal-sharing problems for the students, paying special attention to the number choices and the kinds of fractions and potential equivalence relations that result when students compare different strategies.

Not every child got to share a strategy today. You noticed Gabrielle's strategy in particular, and you plan to ask her tomorrow to use a similar strategy to solve another equal-sharing problem. You decide to write another problem that lends itself to dealing people to brownies. Marina, like other children, had a viable equal-sharing strategy but did not differentiate between halves and wholes. You want to make sure to listen to what she says tomorrow about the fractional quantities she creates.

CONCLUSION

The key in fraction instruction is to pose tasks that will elicit a variety of strategies and representations. Equal-sharing tasks are not the only problems that can do that, but many teachers have found them to be a definite source of variety in thinking. The strategies and representations children produce embody many crucial aspects of fractions. When you are teaching, it may help to organize discussions of strategies around a particular aspect, such as fractions as quantities, fractions as ratios, equivalence, or the geometric character of fractions. In the bulk of these discussions, priority should be given to student-generated strategies. The more students are encouraged to contribute the intact products of their own thinking to class discussions, the more likely they are to identify themselves as understanding mathematics— no matter the level of the thinking. Representational tools and language are supports that children produce or that you provide, as they are needed to enhance, clarify, and communicate strategies.

REFERENCES

Empson, Susan B. "Equal Sharing and Shared Meaning: The Development of Fraction Concepts in a First-Grade Classroom." *Cognition and Instruction* 17 (1999): 283–342.

———. "Equal Sharing and the Roots of Fraction Equivalence." *Teaching Children Mathematics* 7 (March 2001): 421–25.

Hiebert, James, Thomas P. Carpenter, Elizabeth Fennema, Karen Fuson, Diana Wearne, Hanlie Murray, Alwyn Olivier, and Piet Human. *Making Sense: Teaching and Learning Mathematics with Understanding*. Portsmouth, N.H.: Heinemann, 1997.

Hunting, Robert. "The Social Origins of Prefraction Knowledge in Three-Year-Olds." In *Early Fraction Learning,* edited by Robert Hunting and Gary Davis, pp. 55–72. New York: Springer-Verlag, 1991.

Pothier, Yvonne, and Daiyo Sawada. "Partitioning: The Emergence of Rational-Number Ideas in Young Children." *Journal for Research in Mathematics Education* 14 (November 1983): 307–17.

Streefland, Leen. *Fractions in Realistic Mathematics Education.* Boston, Mass.: Kluwer, 1991.

5

Using Manipulative Models to Build Number Sense for Addition of Fractions

Kathleen Cramer

Apryl Henry

THE Rational Number Project (RNP) has reported on several long-term teaching experiments concerning the teaching and learning of fractions among fourth- and fifth-grade students (Bezuk and Cramer 1989; Post et al. 1985). A curriculum created for these teaching experiments and then revised on the basis of what was learned reflects the following four beliefs: (1) children's learning about fractions can be optimized through active involvement with multiple concrete models, (2) most children need to use concrete models over extended periods of time to develop mental images needed to think conceptually about fractions, (3) children benefit from opportunities to talk to one another and with their teacher about fraction ideas as they construct their own understandings of fraction as a number, and (4) teaching materials for fractions should focus on the development of conceptual knowledge prior to formal work with symbols and algorithms (Cramer et al. 1997).

A decade of research on the teaching and learning of fractions among fourth and fifth graders has shown us that of the four pedagogical beliefs listed above, the second is the most important. In order to develop fraction sense, most children need extended periods of time with physical models such as fraction circles, Cuisenaire rods, paper folding, and chips. These models allow students to develop mental images for fractions, and these mental images enable students to understand about fraction size. Students can use their understanding of fraction size to operate on fractions in a meaningful way. The multiple models mentioned above were used in RNP teaching experiments. The fraction circle model used in combination with the RNP activities (see fig. 5.1) was the most powerful of the models. During interviews, students consistently referred to fraction circles as the model that helped them order fractions and estimate the reasonableness of fraction operations.

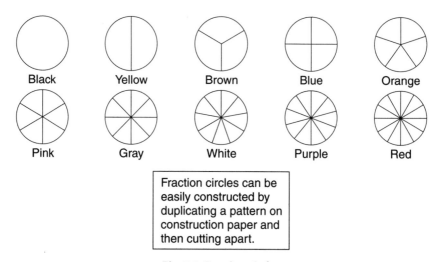

Fraction circles can be easily constructed by duplicating a pattern on construction paper and then cutting apart.

Fig. 5.1. Fraction circles

We know a traditional approach to fraction instruction does not foster students' number sense. Consider the following two fourth-grade students. Both were tracked into the middle mathematics group in a suburban school district south of Minneapolis. They were interviewed the same day as a part of the district's project to field-test the RNP curriculum. The interview was the final one given to both students at the end of five weeks of instruction. Jeremy was in a class using the RNP curriculum, whereas Annie used a traditional textbook program. Jeremy had not learned any procedures for ordering fractions but had interacted daily with fraction manipulatives, exploring fraction concepts and ideas involving ordering at the concrete level. Annie had few opportunities with manipulative materials, since the traditional program moved quickly into procedural skills dealing with fractions. Both children were asked to select the smaller of this pair of fractions: 4/35 and 4/29. Jeremy responded by reasoning, "4/29 is greater. Pieces are bigger, and there's 4 of the smaller pieces and 4 of the bigger piece." Annie reasoned, "4/29 is less. Well, nothing times 29 equals 35, and it's lower numbers."

Annie demonstrated what the RNP documented many times: When students do not have mental images for fractions, they resort to whole-number strategies to try and make sense of the problem (Post et al. 1985). In this instance, Annie's learned procedure did not help her with this problem, since one denominator was not a factor of the other denominator; she therefore used a whole-number ordering strategy. Jeremy, by contrast, was able to reason about the fraction pair. Even though he did not have a concrete model to show 35ths or 29ths, he was able to transfer his thinking about ordering fractions with same numerators—which he did experience concretely in the RNP lessons—to a new situation. We can see that his language is tied to a manipulative model when he states that the "pieces are bigger." During the lessons he used circular pieces of different colors as the main model. Jeremy's thinking is typical of students who have used the RNP curriculum. Table 5.1 shows how he used mental images of fractions in three other ordering tasks from the same interview. Annie's responses to the same questions are also shown. Notice Jeremy's reliance on his mental pictures of the fraction circles.

TABLE 5.1
Other Examples of Students' Ordering Strategies

Problem Task	Jeremy [RNP]	Annie [TEXT]
Order smallest to largest: 1/5, 1/3, and 1/4	1/5, 1/4, 1/3. Pieces for 1/5 are the smallest; for 1/4 there's four to cover the circle; 1/3 there's three.	1/3, 1/4, 1/5. I was thinking of the lowest numbers since all these [numerators] are the same.
5/12 and 3/4: Are they equal or is one less?	3/4 is greater than 5/12. 5/12's are red and not as big. It takes 3 reds to cover a fourth and there's only five of them so it wouldn't cover two of them.	5/12 is less. If you took 4 * 3 to get 12, common denominators and 3 * 3 = 9, it would be 9/12 and it would be more.
4/5 and 9/10: Are they equal or is one less?	9/10 is greater. The pieces are smaller because it takes 10, but there's a smaller gap.	4/5 is less. 5 * 2 = 10 and 4 * 2 = 8 which gives 10/8 and 8 is less than 9.

One purpose of this article is to share with teachers other examples of students' thinking to give them a picture of what it means for students to exhibit number sense with fractions. Interviews with fourth graders who used the RNP curriculum and with fourth-grade students who used a traditional curriculum were conducted by RNP staff as well as classroom teachers. By looking at examples of conceptual and procedural thinking, teachers can construct for themselves a clearer picture of what they will want to observe in their own students. Focusing on children's thinking and on the effect the use of manipulative models has on students' thinking is a way teachers can learn to offer fraction lessons that involve their students with the use of manipulative models over an extended period of time.

ADDITION OF FRACTIONS

A child having fraction sense should be able to estimate a reasonable answer to fraction addition problems. Karen is an example of a fourth grader who was taught the RNP curriculum. She was asked to consider this problem on the day before she worked on fraction addition in her mathematics class (Cramer et al. 1997):

Jon calculated a problem as follows: 2/3 + 1/4 = 3/7. Do you agree?

> *Karen:* I don't agree. He did it weird. You don't add the top and bottom numbers.
> *Teacher:* What would be an estimate?
> *Karen:* It would be… greater than 1/2 because 2/3 is greater than 1/2.
> *Teacher:* Would it be greater or less than 1?
> *Karen:* Less than 1. You'd need 1/3, and 1/4 is less than 1/3.
> *Teacher:* What about 3/7?
> *Karen:* 3/7 is less than 1/2.
> *Teacher:* How do you know?
> *Karen:* Because 3/7 isn't 1/2. I just know.

Notice how Karen, in thinking through this problem, used her knowledge of the relative size of fractions to bring meaning to the problem. She knew 2/3 was greater than 1/2. She did not do any symbolic procedure to compare fractions. She may have known from her many experiences with models for fractions that 2/3 is greater than 1/2. She also ordered 1/3 and 1/4. She realized that 2/3 + 1/3 would equal 1, and since 1/4 is less than 1/3, the total would have to be less than 1. She was less clear about why 3/7 was less than 1/2, but she was sure that was true.

The RNP has documented how children who have had extended periods of time using manipulative models for fractions—and in particular, a circular model—construct for themselves informal ordering strategies to compare fractions (Post et al. 1985). We are using the term *informal strategies* to contrast with traditional ordering procedures of finding decimal equivalents or changing both fractions to equivalent fractions with like denominators. The traditional ordering algorithm is not as helpful in estimation problems as students' constructed strategies.

Karen used two different ordering strategies that the RNP has documented and which reflect students' use of mental images to judge the relative size of fractions (Bezuk and Cramer 1989). She used the same-numerator and the transitive strategies. In the earlier example, Jeremy demonstrated the same-numerator strategy, which involves coordinating an inverse relationship between the size of the denominator and the size of the fraction. In concrete terms, students realize that the more equal parts a unit is partitioned into, the smaller each part is. Jeremy noted that since the numerator was 4 in both fractions, then four of the bigger parts would be larger than four of the smaller parts. Jeremy's thinking noted the role of both parts of the fraction, numerator and denominator. This is important, since often children who learn this relationship as an abstract rule inappropriately apply this thinking when the numerators are different. Karen also used this strategy to conclude that the sum of 2/3 and 1/4 was less than 1 because 1/4 is less than 1/3.

The transitive strategy involves using a reference point such as 1/2. For example, Karen in another interview compared 5/12 and 3/4. She tried to apply an algorithm she had learned along the way to the problem, but she had difficulty. When asked to just think about the fractions in her mind, she responded, "5/12 is less than 1/2; it's one less; 3/4 is more toward the whole than 1/2." What she meant by "one less" was that it was one red piece of a fraction circle away from 6/12, which is 1/2 (twelve reds equaled one circle). Karen used the transitive strategy when estimating fractions when she said that the sum had to be greater than 1/2, since one of the addends, 2/3, was already greater than 1/2.

The use of the transitive and the same-numerator strategies was common among RNP students. Table 5.2 compares RNP and textbook students' estimation strategies (or lack of strategies) for a fraction addition problem. Students were randomly selected to participate in interviews given by project staff during the five instructional weeks. The problem in table 5.2 was on the final interview.

TABLE 5.2
Examples of Students' Fraction Estimation Strategies for the Number Line Problem

Problem	
Tell me about where 2/3 + 1/6 would be on this number line:	0 1 2

Responses	
RNP Students	**Textbook Students**
Two browns are more than a half … 1/6 is not much more … less one.	2/3 equals 4/6 … 4/6 is less than 1/2, so 2/3 + 1/6 is less than 1/2.
Between 1/2 and 1, closer to 1 … 2/3 is almost one; 1/6 is not equal to 1/3, it is less than 1/3.	I have no idea.

Table 5.2 (cont.)

RNP Students	Textbook Students
Between 1/2 and 1 ... 2/3 is a little more than 1/2 ... 2/6 equals 1/3.	Student wouldn't estimate. He tried to find the exact answer. He was incorrect and used that as his estimate.
2/3 is greater than 1/2, 1/6 is a little more ... you need 1/6 more to make a real whole.	Between 1 1/2 and 2 ... just a guess.
Between 1/2 and 1 ... 2/3 is more than 1/2; then add 1/6. It's not greater than 1 because 1/6 is less than 1/3.	Maybe one ... I don't know.
Not a whole ... there's 2/3, and 2/6 is not enough to make 1/3.	Between 1/2 and 1, closer to 1; just a guess.
2/6 fits over 1/3, and there are 2 of them, and you add 1/6 more ... 5/6.	Student wouldn't estimate. He found the exact answer correctly and used that as his estimate.

Despite the same amount of instructional time and an emphasis on teaching the operations, textbook students did not have an adequate understanding of fraction size to find a reasonable estimate to a simple problem. RNP students' thinking depended on mental images for fractions and was directly related to their use of fraction circles. RNP students were better able to verbalize their thoughts. But then, the RNP lessons emphasized students' discussions (belief 3). In the lessons the manipulative model became the focal point of the discussion; students talked about their actions with the fraction circles. The extended use of manipulatives fostered students' verbal skills.

When estimating the sum of 2/3 and 1/6, RNP students reflected on how far 2/3 is from 1. Often RNP students considered the residual, or "leftover part," when they added fractions or when they ordered particular fraction pairs. The power the fraction circle model has on students' images for fractions is particularly evident in this type of reasoning. Consider Jeremy's use of residuals when he ordered 4/5 and 9/10 (see table 5.1). He imagined fraction circles and saw that both fractions had one part missing to make a whole. Since 1/10 is smaller than 1/5, less is missing for 9/10; thus it is the bigger fraction. From this reasoning he knew that 9/10 was greater than 4/5. This type of reasoning required Jeremy to work through several steps mentally. His strong mental images were developed through extensive experiences with fraction circles and supported this multistep thought process.

Since the first RNP teaching experiment, we have documented this type of thinking among fourth graders using fraction circles (Cramer, Post, del Mas in press). As part of the field testing of the RNP curriculum, each classroom teacher interviewed two or three students at the end of the five weeks of instruction. One of the questions was to order 4/5 and 11/12. An example of a residual strategy for this problem is shown by an RNP student. She stated that 4/5 was less than 11/12. Her reasoning was as follows: "4/5 has bigger pieces, but one piece is left; 11/12 has smaller pieces, but only one piece is left. Fifth piece that is missing is bigger." The RNP curriculum did not directly teach students to order fractions by looking at the residual part, nor did it teach students to find common denominators. Results showed that 63 percent of the 53 RNP students interviewed ordered the pair correctly; 21 out of the 53 RNP students constructed for themselves an ordering strategy that relied on residuals. Only 3 of the 53 RNP students used a standard algorithm for ordering these fractions. A few correct solutions relied on drawings. Of the 57 textbook students interviewed, 67 percent ordered the fraction pair correctly. Only 9 out of 57 textbook students used a residual strategy to order these fractions; 18 out of 57 used a cross-product or common denominator method. A few used pictures and found the correct answer but could not explain their reasoning.

CONCLUSION

Often you hear teachers argue that there is not enough time to use manipulative materials. Even when manipulatives are used, teachers often make the transition to symbols too soon. The RNP students discussed in this article used manipulative models virtually every day during five weeks of instruction. The predominant model used was fraction circles. The samples of students' thinking presented here show the benefits of using manipulative models for five instructional weeks. RNP students developed number sense for fractions. In general, they had an understanding of fraction size evidenced by the type of ordering strategies they comfortably used. They were able to estimate reasonable answers to fraction addition problems. They were also able to verbalize their thinking. Students using a traditional program did not develop number sense.

Developing an understanding of fraction size and estimating a reasonable answer to fraction operation problems are appropriate goals for elementary school–aged children. Much of the symbolic manipulation of fraction symbols done in fourth and fifth grades can be adequately addressed in the middle grades. Students will be more successful if teachers in elementary school invest their time building meaning for fractions using concrete models and emphasizing concepts, informal ordering strategies, and estimation.

REFERENCES

Bezuk, Nadine, and Kathleen Cramer. "Teaching about Fractions: What, When, and How?" In *New Directions for Elementary School Mathematics,* 1989 Yearbook of the National Council of Teachers of Mathematics (NCTM), edited by Paul R. Trafton, pp. 156–67. Reston, Va.: NCTM, 1989.

Cramer, Kathleen, Merlyn J. Behr, Richard Lesh, and Thomas Post. *The Rational Number Project Fraction Lessons: Level 1 and Level 2.* Dubuque, Iowa: Kendall/Hunt Publishing Co., 1997.

Cramer, Kathleen, Thomas R. Post, and Robert del Mas. "Initial Fraction Learning of Fourth and Fifth Graders Using Commercial Curricula or the Rational Number Project Curriculum." *Journal for Research in Mathematics Education,* in press.

Post, Thomas R., Ipke Wachsmuth, Richard Lesh, and Merlyn J. Behr. "Order and Equivalence of Rational Numbers: A Cognitive Analysis." *Journal for Research in Mathematics Education* 16 (January 1985): 18–36.

6

Young Children's Growing Understanding of Fraction Ideas

Elena P. Steencken

Carolyn A. Maher

CAN young children learn to understand fractions? This was a question that led us to conduct a yearlong teaching experiment with a fourth-grade class in a New Jersey school. The twenty-five children were heterogeneously grouped, representing a wide range of abilities. They worked on the problems two or three times a week, beginning in late September. The classes were 60 to 80 minutes in duration. We shall report on the first seven of approximately 25 classroom sessions, from September through December, that focused on investigations involving fractions. For all these seven sessions, the children had available Cuisenaire Rods™ to build models. The children worked in pairs (or in small groups) and then came together as a whole class for sharing in larger discussions.

Our team's goal was to pose tasks to the children and to study how their thinking developed by analyzing the videotapes of each session, by studying the children's written work and researchers' field notes, and by listening to students' conversations as they discussed their ideas with one another. Decisions about follow-up activities were based on our best estimates of the children's progress. We also invited children to revisit the same or equivalent problems days or even weeks later. Because we wanted the children to be responsible for determining the reasonableness of their ideas, we encouraged them to decide on the correctness of their solutions and to think about making appropriate generalizations. Our expectations were that each child, or group of children,

This work was supported, in part, by grant MDR 9053597 from the National Science Foundation and by grant 93-992022-8001 from the N.J. Department of Higher Education. Any opinions, findings, and conclusions or recommendations expressed in this paper are those of the authors and do not necessarily reflect the views of the funding agencies.

would construct models of their solutions, and that they would compare these models and their ideas with others. We encouraged them to move about the room and talk to one another. We asked them to prepare their solutions for class sharing and provided overhead rods, transparencies, and overhead projector pens in assorted colors. We worked to promote a classroom culture that supported individual initiative, explanation, and reflection.

In the description that follows, T1 and T2 represent the teacher-researchers. We will use symbols to refer to color and length of each rod: W – white (1 cm), R – red (2 cm), LG – light green (3 cm), P – purple (4 cm), Y – yellow (5 cm), DG – dark green (6 cm), BK – black (7 cm), BR – brown (8 cm), B – blue (9 cm), O – orange (10 cm) (See fig. 6.1). More rod activities are presented in Steencken (1999).

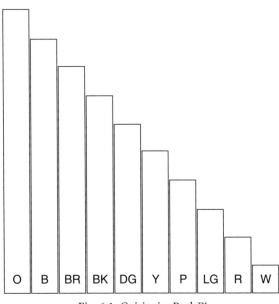

Fig. 6.1. Cuisinaire Rods™

SESSION 1

The first session consisted of a combination of free play with the rods and activities designed to help students distinguish between permanent color names for rods and variable number names. Students were invited to develop ways of supporting their solutions.

Some example activities: "I claim the light green rod is half as long as the dark green rod. What do you think? What would you do to convince me? What number name would we give to the light green rod if I called the dark green rod one? Someone told me that the red rod is half as long as the yellow rod. What do you think?" Students were also invited to pose problems to one another. For example, Alberto challenged, "If the red rod is considered one fifth, what would the orange rod be?" Betty and Mario asked, "If light green is one whole, what is blue? If blue is one, what is light green?

SESSION 2

T1 began by asking the children what occurred in the previous session. Ed began: "Well, let's see. [He chose a blue rod.] If we said that the blue rod would be one whole, we'd figure out what ... we'd take all the blocks and try and figure out what would be one half of it."

The children were invited to find a solution. They concluded that no solution existed for finding a rod that is half as long as the blue rod within the set of rods. Matt and his group explained that for any set of rods, another rod, one half as long, would not exist, that is, the smallest rod in the set.

SESSION 3

The notion of making a "train" of rods—that is, making new rods by placing existing rods end to end—was introduced. Two questions were posed: "If a train of a yellow and a light green rod has the number name *two*, what number name will we give the red rod?"; and "If a train of a yellow and a light green rod has the number name *one*, what number name will we give the red rod?"

Children placed four red rods under their train of yellow and light green rods. Siham and Aleta presented their models using overhead rods and built two identical models with different number names (see fig. 6.2).

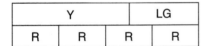

Train of yellow and light green is 2.   Train of yellow and light green is 1.
What is the number name for red?   What is the number name for red?

Fig. 6.2. Students' solution models

The school principal, who frequently visited classes and interacted with the children, asked Aleta to explain the model she built with Siham.

> *Aleta:* Oh, first we put the red rods up to the yellow and light green rods, and then we said, "If the yellow and the green was two, what would we call the red rod?" And we thought that we would call it one and one fourth. And if it [the train] was one, we would call it one fourth.

Principal: Okay, so if it was one, you said you would call it one fourth, and if it was two, what did you say?

Aleta: It would be one and one fourth.

T1 asked the class if they agreed with Aleta's solution. The children indicated that they did not. T1 then asked the girls what they would have to do to convince the class, requesting that the girls first explain the train whose length had the number name *one*. In this train, the girls explained that they named the red rod *one fourth* because there were four red rods whose length together equaled to the length of the train. The class agreed with the girls' name for the red rod when the train was given the number name *one;* but they disagreed with the name for the red rod when the train was given the number name *two*. Bob and Janet pointed out that the red rod should be called *one half*. Dan explained why the red rod should be given the number name *one half:*

> Okay, so this is two [the train given the name *two*], and this would be a half [a red rod] because if you put another one and another one that'd be two. [He aligned four red rods on the overhead.] And if you take away these [two red rods] that would be one and took away that [another red rod], leaving one red rod, that would be a half.

T1 asked the class what they found so confusing about this problem. Ed offered the following explanation:

> I think the confusion is, they think … that they think … they have the temptation of calling … since there are four red blocks, they think they are gonna call it one fourth 'cause they forgot that the yellow and the [light] green are two…. Because, see, if you have one, there'd be two halves; but if you have two it's two halves plus two halves, which would be four halves. Therefore, you would have to call one of the reds one half.

T1 then told the class that she had given to each of two people, Tom and Amy, half of a chocolate bar. Tom was happy with his piece but Amy complained that her amount was unfair. T1 asked how that could have happened, and Mario replied: "You probably gave Tom a bigger half than Amy." T1 asked if that made sense. Meredith responded with an example using yellow and light green rods. She said: "Well, see this was *one* [the yellow and light green train], and then you gave this much to Tom [pointing to the yellow rod, which was five eighths] and this much to Amy [pointing to the light green rod, which was three eighths], that wouldn't be a fair cut."

Dan said that he agreed with Meredith and pointed out that a half should be "even." T1 continued with the candy-bar metaphor. She displayed a large and a small candy bar and said that she gave half of the large one to Tom and half of the small one to Amy. The children giggled and conceded that it was unfair to talk about halves with different wholes.

Next, a comparison problem involving unit fractions was posed to the class: "Which is larger, one half or one third?" T1 invited the students to comment on the meaning of the problem. Matt responded, "Well, normally, one half is bigger than one third, but if you got a bigger size of candy bar or pizza, and if you get one third of that, then that'd be more than one half of a little pizza."

T1 held up a candy bar [scored in a three by four grid]. She asked the children to consider which was bigger, 1/2 or 1/3, in relation to the candy bar. Children worked on a solution and built various models with their rods. The class ended with children agreeing that 1/2 was bigger than 1/3. When all were asked to give a number name for the difference, Alberto shared a model (see fig. 6.3) in which *one* was represented by a train of an orange and red rod. He said, "We know already, that, that, three reds would make a dark green and if there are two dark greens to make an orange and the red rod, then it would take six red rods to make the orange and the red rod."

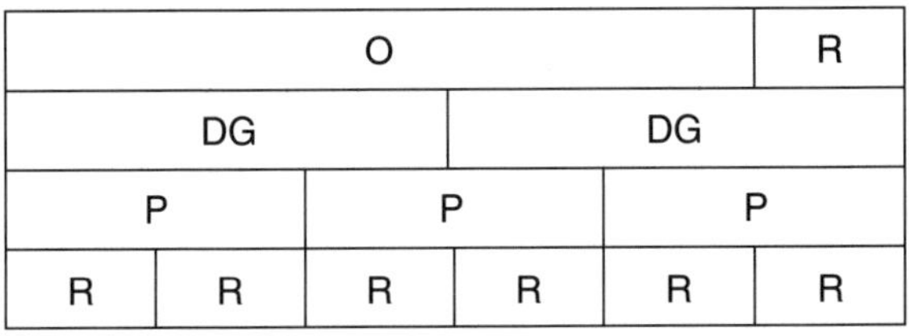

O					R
DG			DG		
P		P		P	
R	R	R	R	R	R

Fig. 6.3. Alberto's model

Various models were built by the children who were convinced that 1/2 was bigger than 1/3; however, they were still working to find a number name for the difference.

SESSION 4

Children were asked to investigate specific relationships between the rods and assign appropriate number names to the rods. T2 asked, "If I call the orange rod one, what number name will I give to the white rod?" The children named the white rod *one tenth*. T2 continued, "If I call the orange rod one, what number name will I give to the red rod?" The children named the red rod *one fifth*. T2 then asked, "If I call the orange rod one, what number name will I give to two white rods?" Mario and Art answered, "One fifth." They went to the overhead to present their solution, which showed the length of the two white rods as equal to the length of the red rod. They gave the red rod the number name *one fifth*.

T2 then asked if there were other answers. In response, Meredith, Betty, Siham, and Dan presented their model, which named the orange rod *one*. The students placed 10 white rods under the orange rod and Meredith and Betty explained:

> *Meredith:* I think it's two tenths. Take the red away and put this [white rod] up to the orange. When we did it before, we said that orange measures ten whites. If you put the whites up it would have ten. Two of ten is two tenths.

> *Betty:* Since ten of these [white rods] equal one orange, then if you took two of these it would be two tenths because one equals one tenth and you just count one more and then you have two tenths.

T2 asked the class how there could be 2 answers, 2/10 and 1/5. Bob showed 2 white rods in one hand and one red rod in the other and explained:

> *Bob:* Even if two white cubes equal up to one red cube, it's still not like imagining that this was another red cube, so I think it's two tenths because it actually is two tenths.

> *T2:* Because you can see two there?

> *Bob:* Yeah, I also see one fifth, but what you're seeing right here is two tenths, not a fifth.

T2 asked the class if it were possible for the two white rods to have both number names. Meredith repeated that the two white rods should be called *two tenths,* and said: "There's only two of them. They're not joined together. If you wanted to join them together you should use a red."

The children were dealing with the issue of equal lengths being represented by physical quantities with different names. Bob and Meredith agreed that the length that Art and Mario called *one fifth* was the same as the length that they called *two tenths.* The distinction was in the number of pieces, "two" for tenths, using white rods; "one" for fifths, using the red rod.

For the remaining 10 minutes of the session, the children were asked to think about the problem from the last session: "Which is bigger, one half or one third, and by how much?" They again agreed that one half was larger than one third. Except for Alberto, they had not spoken of a number name for the difference.

SESSION 5

T1 presented on the overhead the following problem: "Does 1/5 = 2/10?" Meredith came to the overhead and built a model using an orange rod to represent one, and placed two yellow rods beneath the orange rod and explained that the yellow rod would have number name *one half.* She then placed five red rods underneath two yellow rods. She put two white rods together on the screen and explained that two white rods would be named *tenths,* whereas the red rod would be named *one fifth,* indicating that the same lengths can be measured with different units, tenths and fifths.

> *Meredith:* And if you take one of them [She moved one red rod above the two white rods] it is equal to two tenths.
>
> *T1:* So what is your conclusion if I ask you the question, is two tenths equal to one fifth?
>
> *Meredith:* Yes.

Ed and Bob repeated Meredith's explanation to the rest of the class.

T1 produced an overhead transparency of a candy bar [scored in a 3 by 4 grid] and asked the children to give number names for 1/2 of the bar. Janet answered, "... six twelfths, because there are twelve pieces in all, and she got six, and six makes a half." Donna offered, "Two fourths ... because if she got a half, then the top two rows, um, is a half, and then that's two fourths." Bob added, "... three sixths. Yeah, because there are six of them—there, there, there. I found groups of sixths."

T1 then posed the earlier problem: "Which is bigger, one half or one third, and by how much?" Jenny and Lisa volunteered to give their solution at the overhead. They used a train of orange and red rods for *one.* They explained that 1/2 was bigger than 1/3. The teacher asked by how much, and the girls responded that it was a *red* bigger. When asked to give the red rod a number name, Aleta, who joined the other two girls, responded, "It's a third bigger, I think." Jenny and Lisa agreed, and Jenny explained, "I think it's one third bigger, too, because if you put the red to the [dark] green, you need three, and if you put the purple one to it also, and then it takes one third of them." Jenny measured the length of the red rod in comparison to the length of the dark green rod, which was one half in her model. Kate agreed. Bob did not, and said, "Well, I don't really agree. Well, if you split one of the thirds in half, which would make ... which would make a sixth. I think it's a sixth bigger."

T1 asked the class to think about what ideas were being presented. Janet offered a model in which the dark green rod was named *one,* the light green rod was named *one half* and the red rod was named *one third.* She stated, "the light green [*one half*] is one white bigger. After some discussion, the girls agreed that the name of the white rod was *one sixth*. T1 then returned to the original model, where the train of orange and red rods was named *one*. Jenny held to her answer that the difference [a red rod] would be called *one third*. Bob went to the overhead and disagreed with Jenny, who responded, "Well, I think they might both be answers [*one sixth* and *one third*]." When asked, some children agreed with Jenny, some did not. Ed questioned, "How can you have one half be bigger than the thirds by one third? It has to be one sixth." Class ended as T1 asked the children to write about the different arguments that were presented.

SESSION 6

The session began with further discussion of the magnitude of the difference between 1/2 and 1/3. Some children expressed that they could provide a convincing argument that the difference must be 1/6. Jenny offered a model representing *one*, using a train of an orange and a red rod and showed the difference to be 1/6. Alberto's model made use of a dark green rod to represent *one* and also showed the difference to be 1/6. Jenny challenged Alberto's model, "Like remember, you [T1] said that it can be only one size candy bar, and that's it." Janet interjected, saying, "There can be candy bars of different sizes;" Alberto added, "You just can't switch the candy bar."

T1 asked, "What rod has the number name *one sixth* if we call the orange and red [train] *one*?" Siham answered, "Red." T1 then asked the children to name the white rod in this model. Various answers were suggested: *one sixth* [Juan], *one twelfth* [Bob], and *one tenth* [Lisa]. After a lively discussion, Juan changed his mind and went to the overhead. He placed 12 white rods under the train and explained that each white rod would be called 1/12. The children concurred.

A second problem was posed: "Which is bigger, one half or one fourth, and by how much?" The children built a variety of models. Janet, Alice, and Juan went to the overhead and built a model that used a brown rod as *one*, and two purple rods, each as a half. They placed eight white rods under the purple rods and called a white rod *eighths*. They measured the difference between the purple rod [1/2] and two white rods [1/4] as one white rod. They stated that 1/2 was bigger than 1/4 by 1/8. Meredith came to the overhead and showed the group that they still had room in their model for one more white rod, which would make the difference 2/8, not 1/8. She explained that the difference was 2/8 or 1/4, saying, "Yes. Okay. This is an eighth [one white rod]. It's not one eighth 'cause there's still a space. We're calling that an eighth [the white rod]. If you take another one it, um, it could be bigger by two eighths and, or it could be bigger by one quarter. One quarter or two eighths. It's the only way it could be one half." The children concurred.

SESSION 7

The problem, "Which is bigger, two thirds or one half, and by how much?", was presented. Some children immediately produced a model to show their solution; they were then asked if they could build more than one model. Meredith, who built her models in the order shown in figure 6.4, explained her solution to T2.

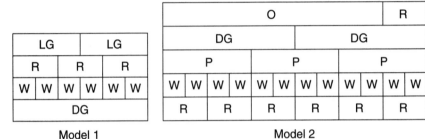

Model 1   Model 2

Fig. 6.4. Meredith's models

Meredith chose two white rods to measure the length between 2/3 and 1/2 in Model 2.

> And then you take the two thirds [she removed two purple rods and one dark green rod from Model 2], and you take two twelfths [the white rods] and then you put it up to two thirds and it is bigger than it by two tens... two twelfths.

Meredith then explained the difference, using Model 1.

> You call these [white rods] sixths, and you put this one [white rod] up to it [one light green rod she removed from Model 1], and it is bigger than [the two red rods she removed from Model 1] one sixth.

T2 asked if there was another number name for the difference between 2/3 and 1/2 in Model 2. Meredith responded.

> Um, yeah, well, maybe ... [Meredith placed six red rods below the train of 12 white rods, and placed a red rod above two of the train of 12 white rods.] ... Yeah, and maybe since two of these little white ones equals up to one of these. [She put one red rod on top of two white rods in the train, showing that a red rod is the same length as a train of two white rods.] Or it's one fifth. Oh, I mean one sixth.

Shortly after this exchange, T1 asked the class to share their ideas with one another. Janet, Ellen, and Jenny presented their solution to the entire class at the overhead and displayed the model that they built. Their model used the same rods for representing *one* as Meredith's Model 2 (see fig. 6.4). They used two dark green rods to show halves, three purple rods to show thirds, and six red rods to show sixths. One red rod was used to show the difference between 1/2 and 2/3. They named the red rod *one sixth*. The teacher researcher asked if there were any other solutions. Meredith responded eagerly and joined the girls. She placed 12 white rods under the six red rods displayed on the overhead. Meredith chose two white rods to measure the difference. She stated that 2/3 was larger than 1/2 by 2/12. Matt strongly objected, "No. ... No, they can't do that. Because the, the two thirds are bigger than a half by a red. So they can't use those whites to show it." T1 questioned Matt's modeling of the problem and asked him for his solution.

> *T1:* And you got two thirds to be bigger than one half?
> *Matt:* [Politely impatient] Yes.
> *T1:* By how much?
> *Matt:* [Deliberately] By one sixth.
> *Meredith:* Or, or two twelfths.
> *Matt:* [Shaking his head in dissent] No. [Mutterings in the classroom of "no"]

Meredith defended her idea and provoked a lively debate among her classmates. Ed offered his understanding of Meredith's solution.

> *Ed:* Yeah, but see just the two whites together. That's right, it would be two twelfths. But you have to combine them. You can't call them, you can call them separately, but you can also call them combined and if you combine them it would be a, a, one sixth.

T1 questioned further.

> *SS:* There's two answers.
> *Matt:* [Simultaneously with Ed] No, they're the same answer.
> *Ed:* No, they're the exact same thing, except she, she took the red and divided it into half, she divided it into halves, into half and called, and called each half one twelfth. They're the exact same answer except they're just in two parts.
> *T1:* So we're talking about the length of the red rod: the length of the red rod is the same as the length of the two white rods? Is that true? Do you all agree to that?

SS: Yeah. Yes.

Ed: And she's calling a white rod one twelfth and the other white rod one twelfth, and the red is really one sixth. Well, when she calls them two twelfths, the two twelfths are actually just two white rods put together to equal a red, so it should be really, it's really one sixth...

Ed offered his understanding of the debate.

Yeah. But I don't really think you could call, call them two twelfths because two twelfths equal exactly to the same size as one sixth. Well, if you want to you could call them, I guess. But I think it would be easier just to call them one sixth cause you wouldn't want to exactly call them one twelfth and another twelfth. I'd just call them one sixth. Therefore I think you just really call them one sixth.

CONCLUSIONS

How did the children's understanding of fractions grow? In the seven sessions, children built some very important understandings about fractions. They built models, drew pictures, and eventually developed notation for their ideas. They began with part-whole relationships and explored patterns created by successive "halving" first with a finite set of rods and later with the idea of infinitely many rods. They focused on the importance of naming the unit in their investigations. In comparing fractions, children developed a firm understanding that units used to represent fractions must be the same size. From rods they moved to other objects such as candy bars and pizzas. Matt commented, "Well, normally, one half is bigger than one third, but if you got a bigger size of candy bar or pizza, and if you get one third of that, then that'd be more than one half of a little pizza." The importance of using the same unit in comparing and finding the difference between fractions was the focal point in their investigations. Alberto reminded us that when comparing fractions you cannot switch the units: "You just can't switch the candy bar."

Children determined the meaning of equivalent number names as they discussed questions such as, "Is two tenths really the same as one fifth?" They explored more possibilities for comparing and naming differences as they built multiple models for their solutions. As the children moved to more-complex problems they made reference often to the rods, even thought they were not building models with them. In discussing their solutions, children listened to one another and developed convincing arguments to support alternative ideas. They eagerly and enthusiastically shared their thinking and provided justifications about the validity of their solutions. They often raised questions and challenges that triggered further exploration and lively conversations. It is clear that young children can build a deep understanding of fraction ideas and enjoy doing it.

REFERENCES

Davis, Robert B. *Discovery in Mathematics: A Text for Teachers.* New Rochelle, N.Y.: Cuisenaire Company of America, 1980.

Maher, Carolyn A. *Alberto's Exploration of Rational Numbers.* Videotape. New Brunswick, N.J.: Robert B. Davis Institute for Learning, 1994a.

———. *The Development of Fourth-Graders' Ideas about Mathematical Proof.* Videotape. New Brunswick, N.J.: Robert B. Davis Institute for Learning 1994b.

———. *Bits and Pieces: Helping Children Build Ideas about Fractions and Proof.* Videotape, in English or Portuguese. New Brunswick, N.J.: Robert B. Davis Institute for Learning, 1995a.

———. *Investigating Ideas about Fractions.* Interactive CD. New Brunswick, N.J.: Robert B. Davis Institute for Learning, 1995b.

Steencken, Elena. *Explorations to Build Meaning about Fractions: Investigation Units for Preservice Teachers,* Vol. 2., edited by Alice Alston. New Brunswick, N.J.: Robert B. Davis Institute for Learning, 1999.

Cuisenaire Rods™ are a registered Trademark of ETA/Cuisenaire®. Permission has been granted to the authors for use of all references to Cuisenaire Rods™ in this article.

Our Rutgers team consisted of Carolyn A. Maher, Amy Martino, the late Robert B. Davis, Elena Steencken, Joan Phillips (a fourth-grade teacher in Conover Road School, Colts Neck, N.J.) and Judith Landis (principal of Conover Road School).

7

Go Ask Alice about Multiplication of Fractions

Susan B. Taber

ALICE'S puzzlement as her size changes in *Alice's Adventures in Wonderland* is similar to the confusion students experience when they are introduced to multiplication with rational numbers. Students find that the operations of multiplication and division have quite different effects from what they did when the multipliers and divisors were whole numbers.

In Wonderland, Alice has no reliable way of predicting whether she will grow larger or smaller. Drinking from the first bottle makes her shrink to ten inches, but drinking from the bottle she finds in the White Rabbit's house makes her grow too large for the house. Eating one cake enlarges her from ten inches to more than nine feet high, but eating the pebble-cakes causes her to shrink small enough so that she can walk through the rabbit's door. Alice begins to be able to control her changes in size after her conversation with the caterpillar (Carroll 1866, p. 68):

> "One side will make you grow taller, and the other side will make you grow shorter."
>
> "One side of *what*? The other side of *what*?" thought Alice to herself.
>
> "Of the mushroom," said the Caterpillar.

Although a round mushroom cap does not have "sides," Alice copes with the situation by breaking off diametrically opposite pieces of the mushroom. After experimenting to find out which part of the mushroom makes her grow and which makes her shrink, Alice is able to change her size to suit her purposes: to participate in the tea party and then to unlock the door and enter the garden. Alice's ability to control whether she will grow larger or smaller corresponds to understanding that multiplication by a number larger than 1 increases the quantity being multiplied, whereas multiplying by a number less than 1 decreases it.

Studies have shown that even after learning to compute the products of fractions or decimals, most students have difficulty solving multiplication word problems that involve fractions or decimals less than 1 (Greer 1992; Graeber 1993). Before learning rules for multiplication of fractions, students learn to find a part of a quantity by division or by subtraction, and they know that fractional quantities are created by dividing or partitioning a quantity into fair shares. Although students learn to compute the products of fractions, many of them continue to think of problems that call for finding a fractional part of the starting quantity as division problems.

Table 7.1 categorizes four types of multiplicative situations and illustrates what occurs when the whole-number factors are replaced by fraction factors. Combine Equal Groups is the most familiar type of multiplicative situation: for example, "There are 6 cars. If each car has 4 wheels, how many wheels are found on the 6 cars?" Multiplicative Compare problems compare two quantities, and Multiplicative Change problems describe the expansion or contraction of a quantity. The two factors in each of the multiplication situations play different roles. One factor is the operator or multiplier; it brings about the change in the other factor, the quantity. Although people easily recognize problems with whole-number multipliers as being multiplication situations (columns 1 and 2 of table 7.1), difficulties arise when the multipliers are fractions or decimals less than 1 (columns 3 and 4 of table 7.1). Many students and adults think that these situations are not multiplication but rather division or subtraction because the resulting quantities are smaller. Furthermore, the familiar Combine Equal Groups problems disappear when a multiplier less than 1 is introduced, and a new kind of problem situation, Partitioning, appears.

TABLE 7.1
Extending Multiplicative Situations to Include Fraction Multipliers and Quantities

Problem Type	Whole-Number Multiplier and Quantity	Whole-Number Multiplier and Fractional Quantity	Fractional Multiplier and Whole-Number Quantity	Fractional Multiplier and Fractional Quantity
Combine Equal Groups	6^1 trays of pies with 3 pies on each tray	6 trays of pies with 2/3 pie on each tray		
Compare	Paul has 3 pies Joe has 6 times as many pies as Paul.	Paul has 2/3 of a pie. Joe has 6 times as much pie as Paul.	Paul has 3 pies. Joe has 3/5 as much pie as Paul.	Paul has 2/3 of a pie. Joe has 3/5 as much pie as Paul.

Table 7.1 (cont.)

Problem Type	Whole-Number Multiplier and Quantity	Whole-Number Multiplier and Fractional Quantity	Fractional Multiplier and Whole-Number Quantity	Fractional Multiplier and Fractional Quantity
Change	Alice is 3 feet tall. After eating the cake, she is <u>6</u> times as tall.	Alice is 2/3 of a foot tall. After eating the cake, she is <u>6</u> times as tall.	Alice is 3 feet tall. After eating the cake, she is <u>3/5</u> as tall.	Alice is 2/3 of a foot tall. After eating the cake, she is <u>3/5</u> as tall.
Partitioning			Paul has 6 pies. He gives <u>3/5</u> of them to Joe.	Paul has 2/3 of a pie. He gives 3/5 of his pie to Joe.

[1]The factor that is the multiplier is underlined.

I collected baseline data from a class of twenty-five fifth graders by asking them to solve a set of word problems like those found in table 7.1. After two weeks of instruction on multiplication of fractions, a majority of the students multiplied to solve problems like this one:

Dave's science report was 1/3 of a page long. Frank's history report was 12 times as long as Dave's report. How many pages long was Frank's report?

The next problem on the page had a fraction multiplier, and most students solved the problem by dividing by the denominator and then multiplying the result by the numerator:

JoAnne's dog weighs 21 pounds on Earth. On a smaller planet it would weigh 2/3 as much. How much would it weigh on the other planet?

Even though students knew how to find the product of a whole number and a fraction, they did not think of the second of these problems as a multiplication problem. Instruction on multiplication of fractions, therefore, needs to help students extend and transform their understanding of whole-number multiplication to include new kinds of problem situations.

Most textbooks introduce multiplication of fractions by demonstrating how finding the product of two fractions solves problems like those in the bottom right-hand corner of table 7.1—those that seem least like multiplication. After presenting the algorithm of finding the product of the numerators and the product of the denominators, the textbooks describe procedures for converting whole numbers and mixed numbers into "improper" fractions so that the algorithm can be used. A more sensible way to introduce multiplicative situations involving quantities and multipliers less than 1 is to begin with situations that are closer to familiar whole-number situations. Even before instruction on multiplication of fractions, many students state that problems involving a whole-number multiplier and a fractional quantity (column 2 of table 7.1) should be solved by multiplication even though they may not know how to perform the operation (Taber 1999). Understanding that multiplication can solve problems like those in the third and fourth columns of table 7.1 requires students to extend and reconceptualize their understanding of multiplication to include multiplication by fraction multipliers.

I designed a fourteen-day instructional unit that focused on understanding the different meanings of problem situations that involve multiplication of fractions for fifth-grade students who had no previous experience with multiplication of fractions. The goals of the unit were as follows:

1. Students would be able to use models and symbolic procedures to find mathematically correct solutions to a variety of multiplication problems having fraction factors.

2. Students would understand that each of the problem types listed in table 7.1 can be solved by multiplication. They would write multiplication number sentences to solve such problems and be able to explain why their answers made sense.

Table 7.1 served as a framework for the instructional unit. Because no one model can adequately represent all the different situations described in table 7.1, I constructed story problems and planned for several physical representations that would give students the opportunity to explore the different problem situations. I planned to introduce problems with whole-number multipliers (like those in column 2 of table 7.1) before introducing problems with fraction multipliers (column 3 of table 7.1). I decided to have students work on problems like those in column 4 of table 7.1 last. I intended to introduce each kind of problem by inviting students to represent and solve problems using physical representations. After students had solved several problems by manipulating the physical representations, I planned to ask them to suggest ways of representing the problems and their solutions symbolically. As part of the process of reaching agreement on appropriate ways to represent problems and compute answers, I intended to ask students to make arguments in support of their suggestions.

During the first two days of the instruction, students represented and discussed Combine Equal Groups problems with fractional quantities like this one: "In our pie shop a piece of pie is equal to three-eighths of a pie. If we sell 6 pieces of pie before lunch, how much pie have we sold?" Students represented the problems with circular fractional pieces, found the answers using their representations, and described what they had done. When we began to discuss how to represent the operation symbolically, we compared these situations with similar situations involving whole-number quantities. We represented 5 × 6 by placing 6 objects in each of 5 bags and 5 × 3/8 by putting 3 one-eighth pieces into each bag. Students agreed that the multiplication symbol should be used to represent the mathematical operation, whether we were multiplying 5 × 6 or 5 × 3/8. After students obtained the answers to several problems by using both their fraction circles and multiplying, I asked them why they were multiplying the numerator by the whole number. Greg explained, "The 8 in the denominator says the kind of part you have. That isn't changing. You are multiplying 5 times 3 of those eighths."

On the third day, we began discussing Partitioning situations with fraction multipliers, such as: "We have 24 balloons. Five-eighths of the balloons are red. How many balloons are red?" Students represented these problems using counters. To help students connect their physical actions of partitioning the set of 24 counters into eighths and collecting five of the eighths with the simultaneous action of the numerator and denominator of the multiplier 5/8 on the starting quantity of 24, I had the students compare this type of problem to multiplication with whole-number multipliers, as in the following discussion:

Teacher: (*Pointing to the written expression* 5 × 20) This number [5] means we are multiplying 20 by—

Student: By five.

Teacher: (*Pointing to* 6 × 5/6) This number [6] means we are multiplying by—

Student: By six.

Teacher: (*Pointing to* 5/8 × 24) Now instead of multiplying something by 5, 6, or 10, or 100, what are we doing?

Student: We're multiplying it by five-eighths.

I next introduced the concept of multiplicative change by showing students a copy of *Alice's Adventures in Wonderland*. We looked at the Tenniel illustrations showing (*a*) Alice holding the little bottle with the label DRINK ME, (*b*) Alice much smaller than the puppy, (*c*) Alice so tall that she can't see her feet, and (*d*) Alice crammed into the White Rabbit's house. All the students had heard of Alice in Wonderland, and from their descriptions of events in the story, it was evident that they had watched the Disney video. They described how eating and drinking made Alice grow and shrink. They reminded me that the small Alice had nearly drowned in the pool of tears that she had cried when she was big.

"Did you know," I asked them, "that Lewis Carroll was really a mathematician named Charles L. Dodgson? Let's see how we can show with mathematics the kinds of things that happened to Alice in the story. Let's suppose that Alice is four and a half feet tall, that's fifty-four inches." I wrote 54 inches on the board and drew a little flask next to it. "Now we're going to have her drink from this little bottle that will make her be one-ninth of her present size," I said as I wrote a 1/9 on the bottle. "How tall will she be after she drinks?"

"Six inches," the students replied. I wrote "= 6" on the board to the right of the bottle.

"That's pretty small," I said. "Suppose it was a one-third bottle she drank from instead." I wrote 54 and drew a bottle labeled 1/3 next to it. "How tall would she be then?"

"Eighteen inches," Jason said, and the rest of the class nodded agreement. I recorded "= 18" on the board.

"Let's have her drink from a—" I began, but was interrupted by Donald, who said, "one-sixth bottle."

I wrote 54 and a bottle labeled 1/6 on the board.

"That's nine inches," Stacey said. I wrote "= 9" on the board.

"Now I want to see what would happen if she drank from a five-sixths bottle," I said, writing 54 followed by a bottle labeled 5/6.

"Oh, this is easy," said several students. "It's 45 inches."

I asked Ken to explain how he got 45 inches. "If one-sixth is nine inches," he said, "then five-sixths would be nine times five, and that's 45."

I called the students' attention to the list we had made on the board (fig. 7.1). "This shows that we're changing Alice's height when she drinks from the bottles. What mathematical symbol should we put in here between Alice's height and the bottles?"

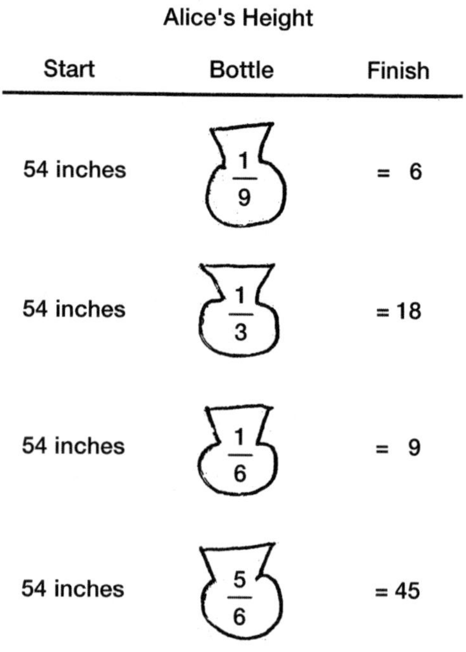

Alice's Height

Start	Bottle	Finish
54 inches	$\frac{1}{9}$	= 6
54 inches	$\frac{1}{3}$	= 18
54 inches	$\frac{1}{6}$	= 9
54 inches	$\frac{5}{6}$	= 45

Fig. 7.1. Charting changes in Alice's size

"Division," Jason said. "You're dividing 54 by one-ninth."

Stephanie raised her hand. "It's multiplication. You're multiplying 54 by one-ninth."

"How many think we're dividing?" I asked. About a third of the class raised their hands. "How many think we're multiplying?" Another third raised their hands. "How many are not sure? Let's take a small Alice and make her get bigger. Maybe that will help us figure out what's going on." We then pretended that Alice ate cakes labeled with different whole numbers and fractions greater than 1. Students computed her height after she ate each cake. We then discussed the appropriate operation symbol to put between the beginning height and either the cake or the bottle. Students agreed that the multiplication symbol would work in either situation—when Alice was shrinking as well as when she was growing.

A student's question before I began the lesson the next day indicated that the students were beginning to understand that multiplication could indeed decrease a quantity. Carrie asked, "Why is it when we times 29 times two-ninths that the answer goes down?"

Wanting to be sure that I understood her question, I asked one of my own: "Your question is that if we multiply 29 times two-ninths, it's smaller than—what?"

"Twenty-nine," Carrie replied. I asked if anyone in the class could think of a reason.

Nina quickly raised her hand, "If you times 29 times 1, you get 29, but two-ninths isn't a complete 1, so you get less than 29."

Alan contributed, "Twenty-nine two-ninths is like counting two-ninths 29 times."

Carrie's question and Nina's and Alan's responses showed that students were beginning to understand the principle of multiplication articulated by the caterpillar in *Alice's Adventures in Wonderland*. They had discovered that multiplying by a number greater than 1 results in a product larger than the starting quantity, but multiplying by a number less than 1 produces a product smaller than the starting quantity. We continued to discuss changes in size that would occur as Alice ate different cakes and drank from a variety of bottles and computed the resulting products. We spent the following two class sessions modeling, discussing, and computing the answers to other Compare and Change problem situations with one fraction factor.

On the seventh day, I introduced problem situations with both fraction multipliers and fraction quantities by referring to the students' explorations of multiplicative change in the context of Alice's changes in size. I asked students how large Alice would be if she drank from a 2/3 bottle when she was only 3/5 of an inch tall. The students agreed that she would be smaller than 3/5 of an inch and that we needed to find a way to compute her new height.

Because students had shown an understanding of using multiplication to find area, I suggested that we represent products of whole numbers and fractions and products of fractions and fractions as areas. We began by representing products such as 2 × 3, 1/3 × 3, and 1/3 × 2/3 on graph paper (see fig. 7.2). Students represented the products on their own pieces of graph paper and then drew their representations on the overhead for the rest of the class.

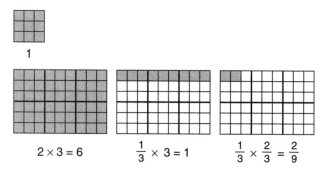

Fig. 7.2. Representing 2 × 3, 1/3 × 3, and 1/3 × 2/3 as areas on graph paper

During the following four class sessions, students represented fraction-times-fraction products as areas and used their drawings to solve the word problems. As we discussed the word problems and students' solutions, the students began to insist that they be allowed to compute the products by multiplying the numerators and the denominators of the fraction factors, since "it is quicker and gives the same answer."

CHANGES IN STUDENTS' THINKING

Before instruction, I had given the students a pretest consisting of twenty word problems, each containing one whole-number and one fraction factor. Half the problems had whole-number multipliers, and half had fraction multipliers. When the multiplier was a whole number, 62.5 percent of the students' strategies attempted multiplication or repeated addition and only 13 percent of their strategies involved division. On problems with a fraction multiplier, however, 59 percent of the students' strategies involved division and none of the strategies attempted multiplication or repeated addition. Interviews with five of the students using four problems like those on the pretest revealed the same pattern of strategy use. Even though each problem contained one whole number and one fraction, students thought about them differently. The students stated that the only problems that could be solved by multiplication or repeated addition were the problems with whole-number multipliers. The students who were able to solve the problems with fraction multipliers used division strategies.

The twenty-problem test was also given as a posttest following instruction. Students' answers indicated that their understanding of multiplication now included situations in which the multiplier is a fraction. On problems with whole-number multipliers, they wrote 97.5 percent of their solutions as multiplication expressions and none as division expressions. On problems with fraction multipliers, 76 percent of the students' strategies involved multiplication and only 21 percent used division.

When I interviewed five of the students the morning before the afternoon posttest, all but one of the students multiplied to solve each of the problems. Their reasons for multiplying on the problems with fraction multipliers provide evidence of their growing understanding that multiplying can partition and reduce quantities as well as combine and enlarge them. For example, Stacy was given a Partitioning problem:

> A carton will hold 42 notebooks. If the carton is 5/6 full, how many notebooks are in the carton?

Stacy said she would solve it by multiplying: "Forty-two times five-sixths."

> *I:* Why do you think it's 42 times five-sixths?
>
> *Stacy:* Because you're going to find it when it's in sixths and then you'll find the five. When you times it and you're going to put it into sixths, it's just like dividing it into sixths and then you get the five sixths.

$$(\textit{She wrote } 42 \times \frac{5}{6} = \frac{210}{6} = 35.)$$

Ed was given this Compare problem with a fraction multiplier:

> A room is 25 feet long. If a rug in the room is 3/7 as long as the room, how long is the rug?

Ed wrote

$$25 \times \frac{3}{7} = \frac{75}{7} = 10\frac{5}{7}$$

and said, "I multiplied 25 times three-sevenths because the room total is 25, and if the rug is three-sevenths as much, so to find out how much three-sevenths of the whole room is, out of the whole room, you multiply so you can find out how much the rug is."

The final interview and the posttest included four additional word problems that involved finding a fractional part of a fractional quantity. The following excerpt from Terry's interview as she solved the first of these problems captures her in the process of establishing for herself that multiplication is an efficient and reliable means of finding a fractional part of a fraction. Here is the problem:

> There was 5/8 of a pizza left over from a party. Tim and Paul each took half of the pizza that was left. How much of a pizza did each boy take?

Terry drew a circle, partitioned it into eighths, shaded five of the sections (fig. 7.3a), and then said, "I think he'd take two pieces and half of one piece."

I[nterviewer]: Can you tell me how much of the whole pizza that would be?

   T: Two and a half eighths.

   I: Can you say that another way?

   T: Five-tenths.

   I: How did you get five-tenths?

   T: I divided all the pieces into halves [*T divides each of the five shaded eighths down the middle* (fig. 7.3b)].

   I: Is that five-tenths of the whole pizza?

   T: No. [*T divides the remaining eighths down the middle* (fig. 7.3c)]. They'd each have five-sixteenths.

   I: So they'd each have five-sixteenths. Could you think of a way to do that with numbers?

   T: Times it by one-eighth.

   I: Show me what you would do.

   T: [*Writes*]

$$\frac{5}{8} \times \frac{1}{2}.$$

Oh! Five-eighths times one-half equals five-sixteenths!

On the next problem Terry multiplied 5/9 times 1/3 to find one-third of 5/9 of a pound of candy. She then multiplied 1/2 times 3/4 and partitioned a drawing to find one-half of 3/4 of a pan of lasagna. She multiplied to solve the remaining interview problem (one-third of 5/6 of a carton of ice cream) and to solve each of the four fraction-of-a-fraction problems on the posttest.

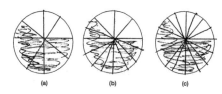

Fig. 7.3. Terry's solutions to 1/2 of 5/8 of a pizza

Even though the teacher, like the caterpillar, can tell students that multiplying by a fraction less than 1 decreases the starting quantity, students, as Alice did, come to understand this concept by experimenting and analyzing the results of their experiments. As a result of their experiences, students in this class reconceptualized their understanding of multiplication to include the actions of shrinking quantities and of finding a fractional part of a quantity.

REFERENCES

Carroll, Lewis. *Alice's Adventures in Wonderland.* New York: D. Appleton & Co., 1866.

Graeber, Anna O. " Misconceptions about Multiplication and Division." *Arithmetic Teacher* 40 (March 1993): 408–11.

Greer, Brian. "Multiplication and Division as Models of Situations." In *Handbook of Research on Mathematics Teaching and Learning,* edited by Douglas Grouws, pp. 276–95. New York: Macmillan Publishing Co., 1992.

Taber, Susan B. "Understanding Multiplication with Fractions: An Analysis of Problem Features and Student Strategies." *Focus on Learning Problems in Mathematics* 21 (Spring 1999): 1–27.

8

Examining Dimensions of Fraction Operation Sense

DeAnn Huinker

Nᴜᴍʙᴇʀ sense and operation sense are similar to problem solving in that they "represent a certain way of thinking rather than a body of knowledge that can be transmitted to others" (Sowder 1992, p. 3). The development of these constructs evolve over time while strengthening intuition about and flexibility with numbers, operations, and their relationships. The need to develop number sense is well documented and its characteristics have received extensive analysis, but the need to develop operation sense has not received the same attention or scrutiny. This discussion presents seven dimensions of operation sense that are applicable to fractions, as well as to whole numbers. These dimensions incorporate a synthesis of the attempts to define and characterize number sense and operation sense (Howden 1989; Sowder 1992). Examples of students' work and reasoning are drawn from work with a class of fifth-grade students from a large, urban school system.

OPERATION SENSE

Fundamental to operation sense is an understanding of the meanings and models of operations. Addition is usually introduced as putting together quantities, and subtraction is presented as separating an amount into subgroups. Multiplication is introduced as combining equal groups, and division is discussed as separating an amount into equal groups. Students should become familiar with a variety of models for fraction operations, such as area models using circular or rectangular regions, set models using counters or cubes, and linear models with number lines.

A class of fifth-grade students was presented with this word problem: Three students are helping the teacher with a special task. He plans to give each student 5/8 of a candy bar to thank them for their help. How much candy will the teacher need so he can give each student 5/8 of a candy bar?

The two students' responses in figure 8.1 show how these students used symbols to model and solve a multiplication situation using repeated addition. Other students used paper-strip models to represent each group demonstrating an understanding of multiplication as combining equal groups.

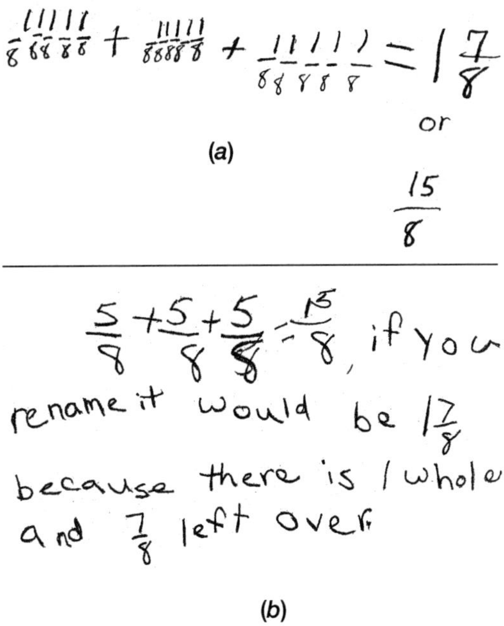

A second dimension of operation sense is the ability to recognize and describe real-world situations for specific operations. Operation sense involves a familiarity with a variety of situations for each operation. For example, addition and subtraction include combining, separating, and comparison situations. Multiplication and division include situations with equal groups, arrays, rates, comparisons, and Cartesian products. Students need to explore many different situations with varied problem structures. They should also regularly describe or pose their own situations for specific operations. The fifth-grade students were presented with a variety of problem structures through daily solving of word problems and were regularly asked to describe real-world situations and pose their own word problems. The word problem in figure 8.2 was posed by a student to illustrate a division situation using fractions.

A third dimension of operation sense involves having meaning for symbols and formal mathematical language. Meaning for symbols and formal mathematical language develops when connections are made to students' conceptual understandings and informal language. Symbols become tools for thinking when students use them as records of actions and things they already know. Without this understanding, students manipulate symbols without meaning rather than thinking of symbols as quantities and actions to be performed or records of actions already performed.

Kieren (1988) noted that premature symbolism leads to symbolic knowledge that students cannot connect to the real world, resulting in virtual elimination of

Fig. 8.1. Two students' approaches for solving $3 \times 5/8$

You have $5\frac{2}{3}$ candy bars.

You had 16 people you wanted to give each $\frac{2}{3}$ of a candy bar. How many people didn't get the candy bar?

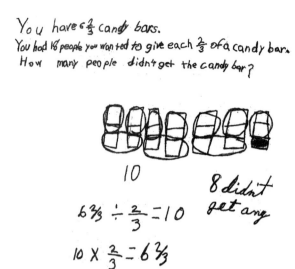

10

$6\frac{2}{3} \div \frac{2}{3} = 10$ 8 didn't get any

$10 \times \frac{2}{3} = 6\frac{2}{3}$

Fig. 8.2. Student-posed word problem for division with fractions

any possibility for students to develop number and operation sense. Symbolic knowledge that is not based on understanding is "highly dependent on memory and subject to deterioration" (p. 178). Thus, operation sense involves having meaning for symbols that contributes to the robustness of knowledge.

Operation sense is strengthened through an emphasis on connecting real-world, oral language, concrete, pictorial, and symbolic representations of fractions. Thus a fourth dimension of operation sense is the ability to translate easily among these representations. Given any one of these situations, students should be able to produce the others. Students who are able to make these connections have demonstrated lasting ability to use their mathematical knowledge flexibly to solve problems. Students who can easily translate among these representations are more likely to reason with fraction symbols as quantities and not as two whole numbers when solving word problems (Towsley 1989).

Students who cannot easily connect these representations lack the power to make sense of fraction concepts and operations and to see the usefulness of fractions in the world around them. The instruction with the fifth-grade students continually emphasized translations from real-world situations to concrete, pictorial, and symbolic representations, thus drawing on varied connections to make sense of and solve word problems. After students examined a situation such as that shown in figure 8.2, explored it using models, and shared solution strategies, the class discussed how to translate the real-world representation into appropriate symbolic representations or equations.

A fifth dimension of operation sense is an understanding of relationships among operations. Addition and subtraction are inverse operations, as are multiplication and division. Multiplication can be viewed as repeated addition, and division can be viewed as repeated subtraction or repeated addition. The use of these relationships contribute to the development of thinking strategies for basic facts and student-generated algorithms for computation. In figure 8.2, note that the student wrote two number sentences. This demonstrates some understanding of the inverse relationship between multiplication and division of fractions.

Operation sense provides a basis for the development of student-generated strategies for computation. Implicit in these strategies is the ability to take numbers apart and put them back together flexibly or to create an equivalent problem. Also implicit in student-generated strategies is the application of the commutative, associative, and distributive properties (Schifter 1999). Thus, a sixth dimension of operation sense involves the ability to compose and decompose numbers and to use properties of operations. For example, Shontae applied the distributive property to solve 2 1/2 ÷ 4. She used two and one half paper strips to solve $(2 \div 4) + (1/2 \div 4)$.

> *Teacher:* You have 2 1/2 candy bars, and there are four people. You are going to share the candy bars so that each person gets the same amount. How much of a whole candy bar would each person get?
>
> *Shontae:* There would be four 1/2 pieces for everyone. Then I thought how many—well, there's still 1/2 left over, and I need to divide it into four pieces. Then I thought that 4/8 is the same as 1/2. So I made it into four pieces, and then if I gave everyone 1/8, they would all have 5/8.

A seventh dimension of operation sense is knowledge of the effects of an operation on a pair of numbers. Operation sense interacts with number sense and enables students to make thoughtful decisions about the reasonableness of results. Understanding an operation includes being able to reason about the effect it will have on the numbers to which it is applied. When you add two numbers, does the answer get larger or smaller? When you subtract, what can you say for sure about the answer? Does the answer always get larger when you multiply two numbers? Does the answer always get smaller when you divide? Can you subtract a larger number from a smaller number? Can you divide a smaller number by a larger number?

Individuals develop expectations about the results of operations from their work with whole numbers. Operating with fractions requires these expectations to be reconsidered. No longer does the answer always get smaller when you divide or larger when you multiply. The students in this class developed an understanding of multiplication and division with fractions through the exploration of problem situations. They reported the results of their explorations in the context of situations that gave meaning to the quantities and relationships among the quantities. Thus they readily reconsidered their expectations regarding the effects of an operation on a pair of numbers and were able to accept and describe, for example, situations in which the answer from a division situation was larger (see fig. 8.2).

Helen, a fifth-grade student, posed the word problem in figure 8.3 for her classmates to solve. She and the other students meaningfully explored the division of a smaller number by a larger number. The response in part *a* shows how a student partitioned each candy bar into eight sections and then gave two pieces to each person. The response in part *b* shows how a student realized that each candy bar only needed to be partitioned into four sections. Because the computation was embedded in a problem situation, the students readily accepted the division of a smaller number by a larger number. During whole-class discussions, the teacher explicitly found opportunities to challenge students' preconceived notions about the effects of operations on a pair of numbers. For example, in the situation shown in figure 8.3, the students had to justify and prove the accuracy of the number sentence, 2 ÷ 8, that they had written to represent the situation.

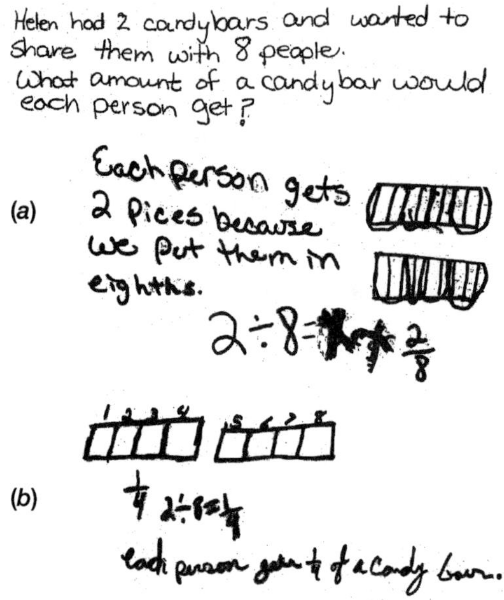

Helen had 2 candybars and wanted to share them with 8 people. What amount of a candybar would each person get?

(a) Each person gets 2 pices because we put them in eighths.

$$2 \div 8 = \frac{2}{8}$$

(b) $2 \div 8 = \frac{1}{4}$

each person gets $\frac{1}{4}$ of a candy bar.

Fig. 8.3. Two students' approaches for solving 2 ÷ 8

CONCLUDING COMMENTS

The work with the fifth-grade students on developing meaning for fraction operations deliberately emphasized the dimensions of operation sense. Instruction used a problem-solving approach and encouraged students to explore operations on fractions with models. The students were encouraged to generate their own strategies for solving problems and to communicate their reasoning by sharing their approaches with one another.

Each student was individually interviewed prior to and following a four-week instructional unit on fractions. The students explored addition, subtraction, multiplication, and division situations with the goal of developing operation sense. Due to time constraints, not all students were asked all items. Table 8.1 lists selected items from the interviews. The preassessments

TABLE 8.1
Interview results with fifth-grade students (percent correct)

Topic	Preassessment Item	Postassessment Item	Number of Students	Preassessment Results	Postassessment Results
Add related fractions	3/4 + 7/8	1/2 + 3/4	25	0%	84%
Subtract a fraction from a whole number	3 – 1/4	3 – 5/8	25	4%	92%
Divide a whole number by a fraction	Not assessed	2 ÷ 2/3	19	----	89%
Divide a mixed number by a whole number	Not assessed	2 1/2 ÷ 4	15	----	60%

revealed that the fifth-grade students overwhelmingly manipulated symbolic representations with little understanding. Some students tried to recall previously learned symbolic procedures but were unable to do so successfully. Students made significant progress in their fraction operation sense by the end of the four weeks. Even though the students were not taught procedures for solving fraction computation problems, the students could draw on their operation sense for fractions. The lower results for 2 1/2 ÷ 4 reflect both the difficulty of this item and the limited instructional time that focused on situations involving division of a mixed number by a whole number.

Fractions are a major area of study in upper elementary school mathematics. The traditional instructional approach for fractions has been heavily symbolic and procedural. The rush to tell students how to perform procedures prevents them from establishing a solid foundation of operation sense for fractions (Kamii 1999). It is time to shift the emphasis and redefine the goal of fraction instruction in elementary school from learning computational rules to developing fraction operation sense. Students with richly connected knowledge of fraction operations are able to develop flexible student-generated strategies for computation and work with problem situations meaningfully.

REFERENCES

Howden, Hilde. "Teaching Number Sense." *Arithmetic Teacher* 36 (February 1989): 6–11.

Kamii, Constance, and Mary Ann Warrington. "Teaching Fractions: Fostering Children's Own Reasoning." In *Developing Mathematical Reasoning in Grades K–12,* 1999 Yearbook of the National Council of Teachers of Mathematics (NCTM), edited by Lee V. Stiff, pp. 82–92. Reston, Va.: NCTM, 1999.

Kieren, Thomas E. "Personal Knowledge of Rational Numbers." In *Number Concepts and Operations in the Middle Grades,* edited by James Hiebert and Merlyn Behr, pp. 162–81. Hillsdale, N.J.: Lawrence Erlbaum Associates; Reston, Va.: National Council of Teachers of Mathematics, 1988.

Schifter, Deborah. "Reasoning about Operations: Early Algebraic Thinking in Grades K–6." In *Developing Mathematical Reasoning in Grades K–12,* 1999 Yearbook of the National Council of Teachers of Mathematics (NCTM), edited by Lee V. Stiff, pp. 62–81. Reston, Va.: NCTM, 1999.

Sowder, Judith T. "Making Sense of Numbers in School Mathematics." In *Analysis of Arithmetic for Mathematics Teaching,* edited by Gaea Leinhardt, Ralph Putnam, and Rosemary A. Hattrup, pp. 1–51. Hillsdale, N.J.: Lawrence Erlbaum Associates, 1992.

Towsley, Ann. "The Use of Conceptual and Procedural Knowledge in the Learning of Concepts and Multiplication of Fractions in Grades 4 and 5." Ph. D. diss., University of Michigan, 1989.

9

Part–Whole Comparisons with Unitizing

Susan J. Lamon

PEOPLE naturally "chunk" quantities differently in their heads. For example, there are many ways in which to think about a case of cola. A case of cola might be 24 (single cans), 2 (12-packs), 4 (6-packs), 1 (24-carton), or 3 (8-bottle packs). Children either chunk information into pieces that are easier for them to think about, or, when discussing everyday items that are in their experience or the packaging they see at home. In fraction instruction, this simple, natural, psychological process often remains covert and causes communication problems.

The solution is not to try to mold children's thinking so that they are all working with the same size chunks. In fact, it is the ability to reconceptualize quantities in different chunks that adds flexibility and usefulness to one's knowledge. For example, suppose we are comparison pricing in a supermarket (see fig. 9.1).

Many textbooks teach children the unit method of comparing prices: find the cost of one ounce

Fig. 9.1. A typical cost-comparison problem

and then multiply to find the cost of the number of ounces you need. After working though a time-consuming long division and then a two-digit multiplication, most children reach the wrong conclusion. Some truncate the first decimal, multiply the resulting error by a factor of 12, and then conclude that the cereals are an equally good buy.

$$16\overline{)3.96} = 0.2475 \approx 0.24 \qquad 12\left(0.24\right) = \$2.88$$

Others round the cost of the Crinkles up to $0.25 per ounce, multiply by 12, and happen to reach the correct conclusion using a procedure that will not always work.

$$16\overline{)3.96} = 0.2475 \approx 0.25 \qquad 12\left(0.25\right) = \$3.00$$

The nuances of truncating and rounding elude young children (and many adults as well), with the result that they end up practicing faulty procedures. In addition, the unit method tends to be a classroom exercise rather than a real-life strategy for comparing prices.

Not many people can do a long division in their heads, and in the supermarket, unless you have a pencil or a calculator handy, the unit strategy is not very helpful. In fact, adults have figured out many other ways to compare prices. One strategy is to compare 4 ounces of each cereal.

For the Crinkles, each set of 4 ounces costs $3.96 ÷ 4 = $0.99.
For the Chunks, each set of 4 ounces costs $2.88 ÷ 3 = $0.96.

Another strategy is to halve the cost of the Crinkles to get the cost of 8 ounces, and halve again to get the cost of 4 ounces:

8 ounces of Crinkles cost $3.98 ÷ 2 = $1.98
4 ounces cost $1.98 ÷ 2 = $0.99
So 12 ounces cost a few cents less than $3.00.

The point is that fraction instruction should take advantage of the natural inclination to chunk quantities, build on it, and encourage flexibility. The process of mentally constructing different-sized chunks in terms of which to think about a given commodity is known as *unitizing*.

Although the given unit in a fraction problem remains unchanged, it may be unitized or chunked in many different ways. For example, represent 3/5 using a rectangular area (see fig. 9.2).

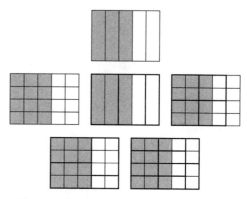

Fig. 9.2. Chunking an area into different-sized pieces

Three fifths is 3 out of 5 columns; it is 12 out of 20 small squares. Other possibilities exist:

$$\frac{3 \text{ (columns)}}{5 \text{ (columns)}} = \frac{12 \ (1 \times 1 \text{ squares})}{20 \ (1 \times 1 \text{ squares})} = \frac{1\frac{1}{2} \ (1 \times 2 \text{ rectangles})}{2\frac{1}{2} \ (1 \times 2 \text{ rectangles})} =$$

$$\frac{3 \ (2 \times 2 \text{ squares})}{5 \ (2 \times 2 \text{ squares})} = \frac{2 \ (2 \times 3 \text{ rectangles})}{3\frac{1}{3} \ (2 \times 3 \text{ rectangles})} = \frac{6 \ (1 \times 2 \text{ rectangles})}{10 \ (1 \times 2 \text{ rectangles})}$$

Although the unit area never changed and the area covered by 3/5 never changed, we could generate different names for 3/5 merely by changing the size of the chunks into which we divided the unit. Relaxing the condition that *a* and *b* be integral and allowing complex forms that could be transformed into the form *a/b*, *b* ≠ 0, opens up a world of possibilities for creatively expressing part-whole comparisons. Similarly, we could use a set of discrete objects:

Mrs. Brown bought a dozen eggs and then used 8 of them for baking pies. What part of the dozen did she use?

$$\frac{8 \text{ eggs}}{12 \text{ eggs}} = \frac{2 \ (4\text{-packs})}{3 \ (4\text{-packs})} = \frac{4 \ (\text{pairs})}{6 \ (\text{pairs})} = \frac{1\frac{2}{6} \ (\text{half dozen})}{2 \ (\text{half dozen})} = \frac{1 \ (8\text{-pack})}{1\frac{1}{2} \ (8\text{-pack})}$$

The unitizing process is also reversible. For example, suppose we know that 6 CDs are 2/3 of the CDs in Bob's collection. How many CDs does Bob have (see fig.9.3)?

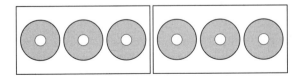

Fig. 9.3. Thinking about CDs in 3-packs

We first notice that 6 CDs is 2 groups of 3.

$$\frac{2}{3} = \frac{2 \text{ (groups of 3)}}{3 \text{ (groups of 3)}}$$

So there must be 3 groups of 3 in the whole collection, or 9 CDs.

There are many advantages in unitizing as an approach to part-whole comparisons. As children enter the world of fractions, there are many cognitive obstacles and points of discontinuity with the whole numbers. The greatest strength of unitizing is that it builds on measurement concepts and children can recognize its consistency with the same measurement principles that they have known since kindergarten. The number of pieces became smaller as we covered the rectangle with larger-sized pieces. The larger the size of the chunks in which we thought about the unit, the smaller the number of pieces we needed to cover the rectangle. Conversely, the smaller the size of chunks in which we thought about the rectangle, the larger the number of the pieces we needed to cover it.

In addition to building on measurement concepts, unitizing affords other advantages:

- Using unitizing to reason up and down replaces the need for rules for generating equivalent fractions as well as for reducing or lowering fractions.
- Unitizing gives children the opportunity to reason about fractions even before they have the physical coordination to be able to draw fractional parts accurately.
- Reasoning up or down while coordinating size and number of pieces lays the groundwork for proportional reasoning.
- Recording the size of the chunks explains how one arrived at the fraction; that is, the notation captures students' thinking so that the teacher can better assess their progress.
- Unitizing appropriately emphasizes a fraction as a number; that is, the emphasis is on the same relative amount, regardless of the size of chunks.
- Unitizing aids self-assessment. Without the use of rules, students can check to see whether they have produced equivalent fractions because the number of chunks multiplied by the number in each chunk never changes.
- Children become comfortable with complex fractions. They are not paralyzed when a problem cannot be solved by halving and doubling, and they have no dependence on perceived patterns or "nice" number relationships to solve problems.

In fact, fourth-grade students who were taught using this approach were able to compare fractions that would typically be beyond their reach until much later in fraction instruction (see fig. 9.4):

Which is larger, 11/12 or 17/18 ?

By the end of fourth grade, students were able to use reasoning, rather than rote procedures, to solve some difficult problems. For example, in the following problem, in which the unit is given implicitly, it is necessary to reason up to the unit and then down to the target fraction. Figure 9.5 illustrates a student's solution.

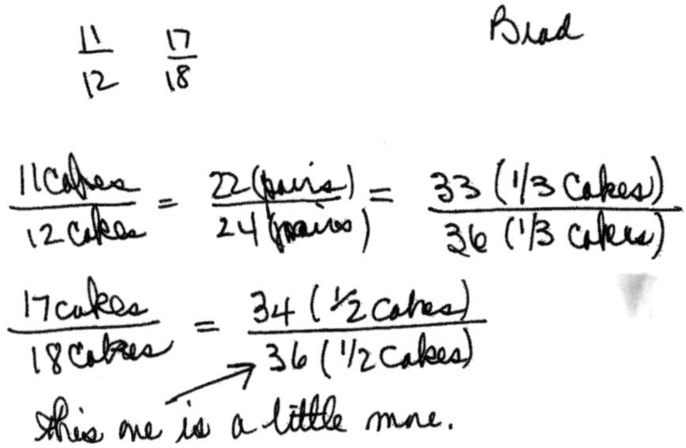

Fig. 9.4. A fourth grader compares fractions by unitizing

I have some baseball cards. Two-thirds of my collection is 6 cards.
I have 5/9 of my collection in my pocket. How many cards are in my pocket?

The students who were taught part-whole fractions with unitizing were never explicitly taught any rules for performing fraction operations. Nevertheless, after a firm foundation in unitizing, they were able to perform fraction operations. For example, fifth graders produced these solutions the first time they were asked to do a fraction division (see fig. 9.6).

Tell how you figure this out: $1 \div 2/3$

These students had clearly built powerful and personal ways of thinking and reasoning. Troy thought about the unit and the divisor in chunks of

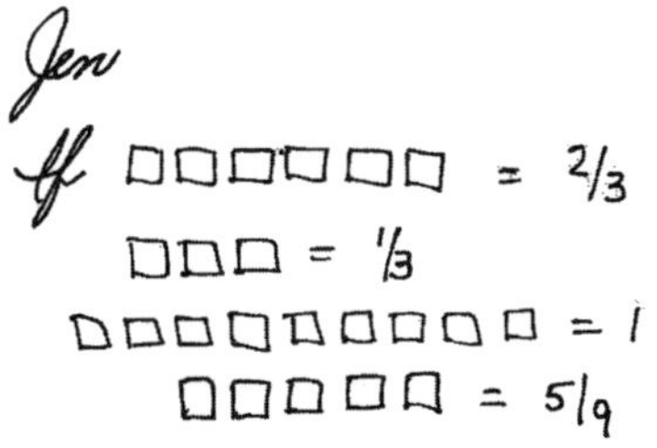

Fig. 9.5. Using reasoning in place of rules and procedures

one-sixth (1/6) units. Then he measured the unit with the divisor to determine that he could take 1 2/4 copies of it. Carson apparently understood 2/3 as the result of dividing 2 copies of the unit into 3 equal shares. He then claimed that starting with only one copy of the unit would yield half as much. Grace knew that 2/3 is twice the size of 1/3. She knew that if she measured with a unit that was twice as large, she would get only half as many copies of it.

What was extraordinary was that, for these students and the others in their class, doing fractions was principally a mental activity. There was some picture drawing, but very little. There were no manipulatives. Merely by thinking, by reasoning up and down, students solved a variety of fraction problems and eventually learned all the fraction operations. In this study, by the end of sixth grade, after four years of studying part-whole fractions with unitizing, there were many more students who used proportional reasoning than can typically be found in most eighth-grade classrooms (Lamon 1999, 2001).

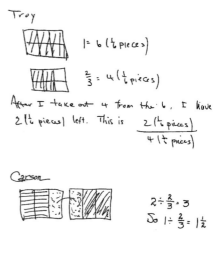

Fig. 9.6. Three solutions to a division problem

With no more than the basic definition of a part-whole fraction, children can begin the unitizing process. First, have students engage in free production activities in which they freely and creatively generate alternate names for a single quantity. For example,

$$17 \text{ shoelaces} = 8\frac{1}{2} \text{ pairs} = 4\frac{1}{4} \text{ (4-packs)} = 34 \text{ (half laces)} =$$

$$170 \left(\frac{1}{10} \text{ laces}\right) = 2\frac{5}{6} \text{ (6-packs)} = 1\frac{7}{10} \text{ (10-packs)}$$

Make sure that they are reasoning up by thinking of smaller-sized chunks, as well as reasoning down by thinking of larger-sized chunks. Then begin to rename fractional parts using the same process, encouraging the free production of multiple names for the fraction.

I have 8 cupcakes. If I eat 2 cupcakes, what part of my cupcakes have I eaten?

$$\frac{2 \text{ cupcakes}}{8 \text{ cupcakes}} = \frac{4 \text{ halves}}{16 \text{ halves}} = \frac{1 \, (2\text{-pack})}{4 \, (2\text{-packs})} = \frac{\frac{1}{2} \, (4\text{-packs})}{2 \, (4\text{-packs})} = \frac{8 \, (\frac{1}{4}\text{-cupcakes})}{32 \, (\frac{1}{4}\text{-cupcakes})}$$

When students have become fluent in generating equivalent fractions by unitizing, introduce a comparison question (see fig. 9.7):

Which amount is greater, 1/2 pizza, or 3/5 pizza?

The length of the string of equivalent fractions will reveal when the student has the idea. If the student is merely generating fraction after fraction, waiting for something interesting to turn up, his or her work may look like Tommy's. As a student gains control over the unitizing process, the string of equivalent fractions generated will become shorter, until, as in Angela's work, you have little doubt that the student has discovered the common denominator and can immediately produce the fractions needed for comparison.

A teacher's patience is required, especially in grades three and four, as students solidify their measurement concepts and learn how to reason with chunked quantities. Restraint is important, too, to ensure that student processes become shorter through understanding and not through rule following. Messy, complex fractions and fractions that are not always in lowest form have to be tolerated. The payoff is worth the wait. Students who develop strong reasoning processes based on unitizing surpass students who have had many years of rule-based instruction, both in their conceptual knowledge and in their ability to perform fraction computation.

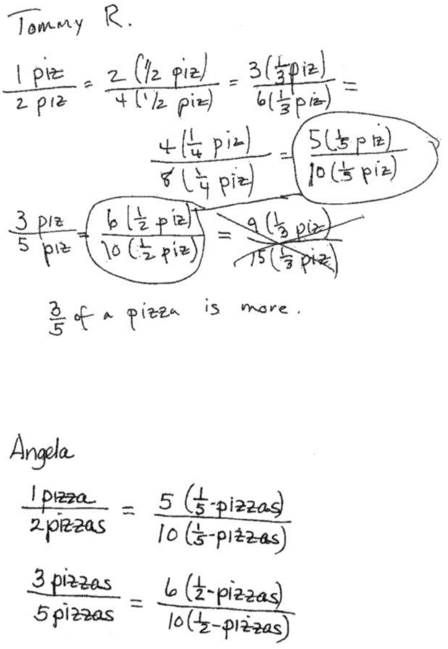

Fig. 9.7. Levels of understanding revealed in children's work on comparison questions

REFERENCES

Lamon, Susan J. *Teaching Fractions and Ratios for Understanding: Essential Content Knowledge and Instructional Strategies for Teachers.* Mahwah, N.J.: Lawrence Erlbaum Associates, 1999.

———. "Presenting and Representing: From Fractions to Rational Numbers." In *The Roles of Representation in School Mathematics,* 2001 Yearbook of the National Council of Teachers of Mathematics (NCTM), edited by Al Cuoco, pp. 146–65. Reston, Va.: NCTM 2001.

10

Butterflies and Caterpillars: Multiplicative and Proportional Reasoning in the Early Grades

Patricia Ann Kenney

Mary M. Lindquist

Cristina L. Heffernan

SUPPOSE you asked fourth-grade students to answer each of the two questions shown in figure 10.1. Would their answers show their understanding of multiplicative or proportional reasoning? What other kinds of reasoning or strategies might they use to answer the question? How would they use pictures, words, and numbers to answer the question? What kinds of errors would you expect to see, and how would you help students to overcome those errors?

The two questions in figure 10.1 are from the 1996 National Assessment of Educational Progress (NAEP) in mathematics given to fourth-grade students. Examining these two NAEP questions and responses from students in the NAEP sample can provide insight into how students in the elementary grades solve problems that involve multiplicative and proportional reasoning. The responses are also illustrative of the strategies and misconceptions that have been documented in several research studies concerning multiplicative and proportional reasoning in the elementary grades.

The preparation of this article was supported in part by a grant to the National Council of Teachers of Mathematics (NCTM) from the National Science Foundation (NSF), grant no. RED-9453189. Any opinions expressed herein are those of the authors and do not necessarily reflect the views of NCTM or NSF.

The children who visit a booth at a science fair are going to build models of butterflies. For each model, they will need the following:

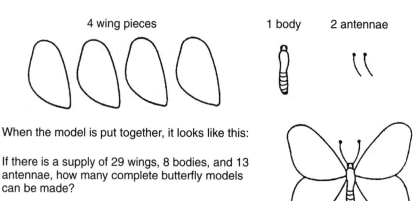

When the model is put together, it looks like this:

If there is a supply of 29 wings, 8 bodies, and 13 antennae, how many complete butterfly models can be made?

Answer: _____

Use drawings, words, or numbers to explain how you got your answer.

(a)

A fourth-grade class needs 5 leaves each day to feed its 2 caterpillars. How many leaves would they need each day for 12 caterpillars?

Answer: _____

Use drawings, words, or numbers to show how you got your answer.

(b)

Fig. 10.1. Two NAEP questions for grade 4

THE BUTTERFLY MODELS AND THE CATERPILLARS AND LEAVES TASKS IN THE NAEP ASSESSMENT

For more than twenty-five years, NAEP has been the only continuing assessment of what American students know and can do in a variety of school subjects, including mathematics. Since 1972, the NAEP mathematics assessment has been given about every four years to a representative sample of students; in recent assessments, the samples have included students in grades 4, 8, and 12. Periodically, the NAEP mathematics assessment includes special studies that focus on particular areas of interest to mathematics educators. For example, the 1996 assessment included a special study of mathematics in context at grades 4, 8, and 12; that is, students' performance on sets of related problems based on a thematic context that was thought to be familiar and interesting to most students at that grade.

One of the theme blocks administered to fourth-grade students was situated in the context of a science fair project on butterflies. On the first page of the problem set a short paragraph introduced the context to the student and was followed by six questions, two of which were (1) Butterfly Models and (2) Caterpillars and Leaves (see fig. 10.1). While working on the butterfly-related questions, all of which were in constructed-response format, students had the use of a four-function calculator, a ruler, and other materials, such as colorful pictures of butterflies and a set of butterfly cutouts.

From work associated with the NCTM-NAEP Interpretive Reports (Kenney and Lindquist 2000), we had the opportunity to study a small sample of students' work on questions from the fourth-grade theme block on butterflies. The sample of 172 students was a convenience sample and was not meant to represent the responses of *all* fourth-grade students in the NAEP sample but instead to provide examples of students' work. We were particularly drawn to the Butterfly Models and Caterpillars and Leaves tasks because of their focus on multiplicative and proportional reasoning, topics that are viewed as "capstones of the elementary school mathematics curriculum" (Lesh, Post, and Behr 1988) but that were not assessed anywhere else in the grade 4 NAEP mathematics assessment. In the next sections, we focus on each problem, examining the ways in which students found the correct answer, their misconceptions that led to incorrect answers, and how some students got the correct answer for the wrong reason.

Butterfly Models

The Butterfly Models problem (see fig. 10.1a) asked students to decide how many complete butterfly models can be made from a given supply of wings, bodies, and antennae. Their work should include a complete explanation about how 6 complete models can be made using 6 bodies, 24 wings, and 12 antennae of the given supply. Because the solution to the task involves a multiple of 6 (6 bodies, 6×2 antennae, 6×4 wings), the task has the potential to reveal students' understanding of *multiplicative reasoning*.

When we examined students' work from the sample set of responses, we found that there were a number of ways that students solved the problem, including considering all three body parts simultaneously or focusing on only one of the parts. Important misconceptions such as the misuse of the additive model and computational errors were also evident in the students' work.

Considering all butterfly model parts

There is evidence in our sample that some students considered the entire supply of butterfly model parts simultaneously by either drawing actual models or using multiplicative or additive reasoning. Responses 2.1 through 2.4 in figure 10.2 are typical of how students used pictures to keep track of the number of parts used and the parts left over. Response 2.1 is from a student who accounted for all extra pieces by correctly drawing six complete models and two incomplete butterfly models, one having four wings and one antennae and the other having one wing and no antennae. In producing response 2.2, the student combined pictures with words, correctly noting that "1 antennae, 5 wings, and 2 bodys" would be left over. Response 2.3 is perhaps not as explicit as the other two, but the words "no more antennae" suggest that this student is basing his or her answer on the lack of enough antennae to make another butterfly. Response 2.4 is the least explicit because we do not know why the student stopped drawing parts after the sixth model.

Although the pictorial strategy—even with words—is one way to solve the problem, it tells us little about how—or if—students were using multiplicative or additive reasoning to solve the problem. Responses 2.5 through 2.8 show this use of multiplicative and additive reasoning more explicitly. The student who produced response 2.5 used a table to combine pictures and a strategy that could be either multiplicative or additive. That is, from the work shown for the rows associated with wings (4, 8, 12, 16, ... 28) and antennae (2, 4, 6, 8, ... 12), we cannot tell if the student was multiplying by 2 ($4 \times 2 = 8$ wings; $2 \times 2 = 4$ antennae) or adding four wings ($4 + 4 = 8$) and 2 antennae ($2 + 2 = 4$) for each model. Response 2.6 shows a similar reasoning strategy that could be either additive or multiplicative but without model building.

Response 2.7 shows an interesting use of a chart with an additive (or subtractive) method, but it lacks a complete explanation. However, we can imagine what the student might have been thinking as he or she produced the table. After subtracting 4 from 29 and getting 25 wings, the student

Answer of 6: Pictorial Models (with or without words)

Fig. 10.2. Examples of responses to the Butterfly Models task that consider all body parts simultaneously

began by "creating" the first model and then worked with the parts remaining. For example, the triple (25, 7, 11) is the number of body parts left after making one model. The subsequent columns in the table are the number of parts remaining as more models are built. The triple (5, 2, 1) in the last (sixth) column shows that although enough wings (5) and bodies (2) are left, only one antenna remains. Using continued subtraction, then, this student did actually answer with 6 models.

Response 2.8 also begins with a calculation, but this time the student uses division to find that 7 models can be made with 29 wings. This student uses words to continue a multiplicative reasoning strategy that leads to a string of statements about the relationship between the number of parts and the number of complete models.

In addition to the strategies shown above that resulted in correct answers, the sample responses also revealed some ways in which students obtained incorrect answers. In our sample set, the most common error among students who considered all butterfly model parts was merely to add the given numbers (29 + 8 + 13) and answer 50 models; this was also the most common answer overall. Some students just added the number of parts required to make one model (4 + 2 + 1) and answered 7 models. In their work with NAEP results, Kouba, Zawojewski, and Strutchens (1997) refer to this misuse of the additive model as the "when in doubt, add" strategy.

Focusing on one butterfly model part

For students who apparently considered only the number of antennae, the most common work shown was either a multiplication or a division problem ($6 \times 2 = 12$ or $12 \div 2 = 6$ or adding 2 six times) or a statement like "There are only enough antennae for 6 butterflies." Just a few students in our sample showed work that led us to assume that they considered only the antennae. It was more likely that students focused on the number of bodies and said "8 models." For example, one student wrote this explanation: "If there was 8 bodies then that means there has to be 8 butterflies because butterflies don't have 2 bodies." However, these students failed to notice that there were not enough antennae for 8 models.

Another group of students focused on the number of wings, and common answers were 6, 7, or 8 models. Some students tended to make an error when dividing 29 wings by 4, answering 6 with a remainder of 1, thus obtaining the correct answer apparently for the wrong reason. Students also wrote that $29 \div 4 = 8$. The answer of 7 models comes from correctly dividing 29 by 4, not realizing that the number of antennae in the supply does not allow for 7 complete models.

Summary

From students' responses in our sample, we found examples of students using a variety of methods to solve the Butterfly Models task, including drawing pictures and using multiplicative and additive reasoning. However, it was often difficult to tell from their work whether students were reasoning multiplicatively or additively. Additionally, most of the students in our sample failed to get the correct answer.

Our findings about the methods students used and the difficulty they had in solving the Butterfly Models task reflect the results from NAEP's national sample of fourth-grade students. NAEP results revealed that this was a difficult problem, since only 3 percent of the students gave a response that gave both the correct answer and a complete explanation. Another 23 percent said 6 models with no explanation, 7 models with a reasonable explanation, or a response that showed some evidence that the number of parts influences the number of models. However, slightly more than 60 percent gave an incorrect answer.

Caterpillars and Leaves

The Caterpillars and Leaves task (see fig. 10.1b) asks students to find the number of leaves needed to feed 12 caterpillars if it takes 5 leaves to feed 2 caterpillars. This task is similar to the kinds of *proportional reasoning* problems commonly found in curricular materials at upper grades. Again, students could use a variety of methods to solve the problem, such as drawings, tables, written explanations, and arithmetic methods. Responses also showed errors and misconceptions about proportional reasoning that are well documented in the research literature, such as thinking about proportion in an additive way and misunderstanding or not recognizing a proportional situation.

Answers of 30 leaves with correct explanation

As with the Butterfly Models task, we noticed that students, in explaining a response of "30 leaves," coupled pictures with words or charts and showed evidence of additive, multiplicative, and proportional reasoning. Responses 3.1 through 3.7 in figure 10.3 are typical of the kinds of responses we found in the sample. The students who produced responses 3.1 and 3.2 reasoned using a single unit rate for the number of leaves for each caterpillar; that is, these students noted that each caterpillar needed 2 1/2 leaves each day and then used this value to solve the problem. A pictorial model is used in response 3.1, but we do not know if the student used addition or multiplication to arrive at the answer; in response 3.2 there is little doubt about the use of multiplication ($2\,1/2 \times 12 = 30$). In this example, once the student had calculated the number of leaves for each caterpillar, this number could be multiplied by 12 to get the number for 12 caterpillars.

Response 3.3 shows a similar strategy to that of the single unit rate. The difference is that in response 3.3 the student uses the given relationship of 2 caterpillars to 5 leaves as what Lamon (1995) calls "a composite unit." To solve the problem using this composite unit, the student first finds the multiplicative relationship between 2 caterpillars and 12 caterpillars. Because there were 6 times as many caterpillars, then there should be 6 times as many leaves ($6 \times 5 = 30$). Both the unit rate strategy and the composite unit strategy for proportional reasoning are well documented (Cramer and Post 1993; Hart 1988; Lamon 1995).

Responses 3.4 through 3.7 are based on the strategy of counting up to the number of leaves. Responses 3.4 and 3.5 are examples of a strategy that Lamon (1999) calls "building up." Using a building up strategy results in finding that 12 caterpillars need 30 leaves, as long as no computational errors occur. As with the tables constructed to answer the Butterfly Models task (see fig. 10.2.5), it is unclear whether the students used a multiplicative or an additive strategy to construct them. In response 3.6, a similar strategy was used, but this student recorded the 2 and the 5 each time instead of recording the totals. The totals look as if they were found at the end; however, it is difficult to know how the student knew to stop after six lines. It is possible that the student was keeping track of the total number of caterpillars while recording the data in the table. Response 3.7 is similar to response 3.6 except that this student did not record the work in a table.

Although some students used additive, multiplicative, and proportional reasoning to solve the Caterpillars and Leaves task, we also found evidence that other students could not do this well. Instead, their explanations suggested that they were thinking about proportion only in an additive way or that they were misunderstanding the proportional situation in the problem.

Thinking about proportion in an additive way

An answer of 15 indicates that some students were thinking about the relationship between leaves and caterpillars in an additive way. That is, they thought that the amount to be added must remain constant rather than the multiplier remaining constant. Three examples of this in relation to the Caterpillars and Leaves task appear in figure 10.4. In response 4.1, the student apparently focused on the fact that there were 10 more caterpillars rather than 6 times as many caterpillars. He or she then added 10 more leaves to the original 5 leaves, keeping the amount added constant. Response 4.2 shows this way of thinking but uses a chart to track the increase in leaves by caterpillars. Similarly, in response 4.3 the student most likely reasoned that since at first there were 3 more leaves than caterpillars, then there should be 3 more leaves for the 12 caterpillars, or 15 leaves.

Response 3.1

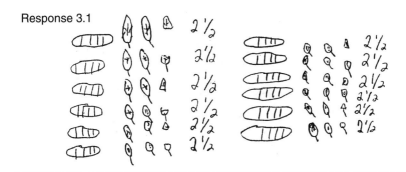

Response 3.2

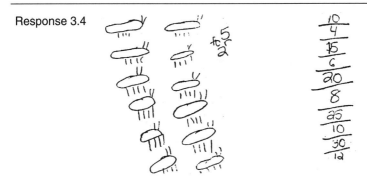

if each of these caterpillars need
2 ½ leaves a day then you
just X's 2½ x's 12 = 30.

Response 3.3

If it takes 5 leafs for two
Caterpillars, you just count
by twos, until you come to
half of 12. The number
is six, and then you
multiply 5 X 6, and it
equals 30.

Response 3.4

Fig. 10.3. Examples of responses to the Caterpillars and Leaves task that show correct answers of 30 leaves along with appropriate explanations

Response 3.5

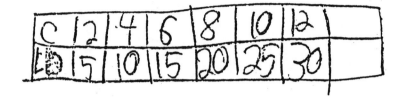

Response 3.6

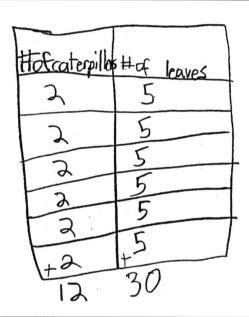

Response 3.7

I knew the answer because
I keeped counting by 2
and keeped -adding by 5's,

Fig. 10.3. (*cont.*)

Response 4.1

They added 10 catarpillers, and so I added 10 leaves.

Response 4.2

leaves	caterpillars
5	2
6	3
7	4
8	5
9	6
10	7
11	8
12	9
13	10
14	11
15	12

Response 4.3

I looked at the numbers 5 and 2 and I added 3+2=5 so I went three more numbers that 12.

Fig. 10.4. Examples of responses to the Caterpillars and Leaves task showing thinking about proportion in an additive way

In studying proportional reasoning, Hart (1988) and Lamon (1995) have found that students have a difficult time losing the instinct toward additive reasoning rather than multiplicative reasoning. This difficulty is attributed to the emphasis on additive reasoning that pervades the elementary school mathematics curriculum, making the transition to multiplicative reasoning a challenge.

Misunderstanding a proportional situation

In our sample of responses to the Caterpillars and Leaves task, the two most common answers by far were 60 leaves and 19 leaves. Each answer involves a misunderstanding of the situation in the problem. An answer of 60 leaves was usually accompanied by an explanation in the form of a multiplication problem: $5 \times 12 = 60$. Perhaps students who gave this answer understood that the problem involved a multiplicative situation, but they did not have a complete understanding of the situation, which called for 2 caterpillars (not 1) for every 5 leaves. As with the Butterfly Models problem, students also tended to add up the numbers given in the problem for no apparent reason $(5 + 2 + 12)$ and therefore answered 19 leaves.

Summary

From students' responses in our sample, we found examples of students reasoning proportionally using strategies such as unit rate, composite unit, and building up with multiplication or addition. Some students also used pictures and tables to help them solve the problems. However, as with the Butterfly Models task, most of the students in our sample failed to get the correct answer. This pattern reflects that of the national sample of fourth-grade students. Again, NAEP results revealed that this was a difficult problem: only 6 percent of the students gave a correct answer with a correct explanation, and another 7 percent either gave a correct answer without an explanation or used a correct method with a computational error (for example, incorrectly multiplying $2\ 1/2 \times 12$). The majority of students in the NAEP sample, 86 percent, gave an incorrect response.

CONCLUSION

The *Principles and Standards for School Mathematics* (NCTM 2000) calls for more emphasis on multiplicative reasoning in grades 3–5. Many different topics in the curriculum at these grades (e.g., place value, area, and volume) depend on multiplicative ideas. Students should understand when multiplication (or division) is an appropriate and efficient strategy for solving a wide range of problems.

Students' work on the two NAEP items show that some students have emerging ideas about multiplicative and proportional reasoning. Although each of these problems could be solved without multiplication or without proportions, students at this age should be moving toward using multiplication (or division) in situations like these. They need the opportunity to solve problems like these but where multiplication becomes essential or where they can use multiplication (or division) to solve proportions. Mistakes such as the ones students made on these two items should be seen as opportunities for learning. Try these items with your students, listen to their approaches, and build on their understandings.

REFERENCES

Cramer, Kathleen, and Thomas Post. "Proportional Reasoning." *Mathematics Teacher* 86 (May 1993): 404–7.

Hart, Kathleen. "Ratio and Proportion." In *Number Concepts and Operations in the Middle Grades,* edited by James Hiebert and Merlyn Behr, pp. 198–279. Reston, Va.: National Council of Teachers of Mathematics, 1988.

Kenney, Patricia Ann, and Mary M. Lindquist. "Students' Performance on Thematically Related NAEP Tasks." In *Results from the Seventh Mathematics Assessment of the National Assessment of Educational Progress,* edited by Edward A. Silver and Patricia Ann Kenney, pp. 343–76. Reston, Va.: National Council of Teachers of Mathematics, 2000.

Kouba, Vicky L., Judith S. Zawojewski, and Marilyn E. Strutchens. "What Do Students Know about Numbers and Operations?" In *Results from the Sixth Mathematics Assessment of the National Assessment of Educational Progress,* edited by Patricia Ann Kenney and Edward A. Silver, pp. 87–140. Reston, Va.: National Council of Teachers of Mathematics, 1997.

Lamon, Susan J. "Ratio and Proportion: Elementary Didactical Phenomenology." In *Providing a Foundation for Teaching Mathematics in the Middle Grades,* edited by Judith T. Sowder and Bonnie P. Schappelle, pp. 167–98. Albany, N.Y.: State University of New York Press, 1995.

———. *Teaching Fractions and Ratios for Understanding: Essential Content Knowledge and Instructional Strategies for Teachers.* Mahwah, N.J.: Lawrence Erlbaum Associates, 1999.

Lesh, Richard, Thomas Post, and Merlyn Behr. "Proportional Reasoning." In *Number Concepts and Operations in the Middle Grades,* edited by James Hiebert and Merlyn Behr, pp. 93–118. Reston, Va.: National Council of Teachers of Mathematics, 1988.

National Council of Teachers of Mathematics (NCTM). *Principles and Standards for School Mathematics.* Reston, Va.: NCTM, 2000.

Classroom Challenge

Helen A. Khoury

Northern Illinois University, DeKalb, Illinois

Exploring Proportional Reasoning: Mr. Tall/Mr. Short

This is Mr. Short:

Fig. 1. An adapted version of the Mr. Tall/Mr. Short problem

The length of Mr. Short is 4 large buttons.
The length of Mr. Tall is 6 large buttons.

When paper clips are used to measure Mr. Short and Mr. Tall:
    The length of Mr. Short is 6 paper clips.
    What is the length of Mr. Tall in paper clips?_____

Please **EXPLAIN** how you arrived at your answer.

Many researchers and teachers have been studying students' proportional reasoning for at least the last thirty years. Robert Karplus and his colleagues (1970, 1974, 1980) are the pioneers among this group of researchers and teachers. The Mr. Tall/Mr. Short problem was designed by Karplus to assess students' levels of proportional thinking. Since then, this problem has been widely used by teachers, teacher educators, researchers, and test developers.

Mr. Tall/Mr. Short is a paper-and-pencil problem (see fig. 1). It is used with students either in classroom or in individual settings to identify and to assess the following levels of proportional thinking:

Level I (illogical): No explanation is given, and an illogical computation, a guess, or a general estimate is made on the basis of the descriptive observation: "Mr. Tall is 10 paper clips, ... because he is tall, so 4 + 6 = 10."

Level A (additive): The student focuses on the difference between 6 and 4 buttons, and then assumes that the same difference needs to exist when the paper clips are used: "Mr. Tall is 8 paper clips, 6 - 4 = 2 ..., and 6 + 2 = 8 paper clips."

Level TR (transitional): The student uses an additive approach that focuses on the correspondence of the measures of each figure: "Mr. Tall is 9 paper clips. Mr. Short is 6 paper clips, 2 more than 4. So, for each 2 buttons there is one more paper clip. The same should hold for Mr. Tall [(2 + 1)+(2 + 1) + (2 + 1)]."

Level R (ratio): The student uses a constant ratio relationship or makes a multiplicative comparison of the measures of both figures: "Mr. Tall is 9 paper clips, ... because for every 1 button for Mr. Short, Mr. Tall is 1 1/2 buttons. When Mr. Short is 6 paper clips, Mr. Tall is 9 paper clips (6 times 1 1/2)."

In analyzing students' work, the teacher needs to focus on students explanations of how they arrived at their answers. Students should be encouraged to use the mode(s) of representation of their choice (pictures, symbols, numerals, written words, tables, and so on) to explain their solution strategies.

References

Karplus, Robert, E. Karplus, and W. Wollman. "Intellectual Development beyond Elementary School IV: Ratio, The Influence of Cognitive Style." *School Science and Mathematics* 74, no. 6 (1974): 476–-82.

Karplus, Robert, and R. Peterson. "Intellectual Development beyond Elementary School II: Ratio, a Survey." *School Science and Mathematics* 70, no. 9 (1970): 813–20.

Karplus, Robert, H. Adi, and A. Lawson. "Intellectual Development beyond Elementary School VIII: Proportional, Probablistic, and Correlational Reasoning." *School Science and Mathematics* 80, no. 8 (1980): 673–83.

Classroom Challenge

Patrick W. Thompson

Vanderbilt University, Nashville, Tennessee

3/5's Problem

The following questions help students see parts and wholes in multiplicative relationships. You, as a teacher, can ask them to see many fractions and relationships. In each example, the student must think carefully about what is the part and what is the whole as it relates to that part. The kind of thinking you want students to engage in is shown in parentheses after each question.

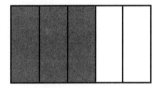

Questions to ask:

Can you see 3/5 of something in this figure? (Think of the whole thing as one. Then think of the shaded part in relation to the whole.)

Can you see 5/3 of something? (Think of the shaded part as one. Then think of the whole thing in relation to the shaded part.)

Can you see 5/3 of 3/5? (Think of the whole thing as one. Then think of the shaded part in relation to the whole, which gives 3/5 of one. Then think of the shaded part as one 3/5. Then think of the whole thing in relation to the shaded part. The

From: Thompson, Patrick W. "Notation, Convention, and Quantity in Elementary Mathematics". In *Providing a Foundation for Teaching Mathematics in the Middle Grades.* edited by Judith T. Sowder and Bonnie P. Schappelle, p. 199–219 Albany, N.Y.: State University of New York Press, 1995.

whole thing [one] is 5/3 of the shaded part, which itself is 3/5 of one. The whole thing [one] can be seen to be 5/3 of [3/5 of one].)

Can you see 2/3 of 3/5? (Think of the whole thing as one. then think of the shaded part in relation to the whole, which gives 3/5 of one. Now think of 2/3 of the shaded part, which is 2/3 [3/5 of one], which is also 2/3 of one. Two-fifths of one can be seen to make up 2/3 of [3/5 of one].)

Can you see $1 \div 3/5$? (Think of the whole thing as one. Then ask, "How many units of 3/5 are in one?" There is one 3/5 and 2/3 of another 3/5. The one whole can be seen as containing 1 2/3 units of [3/5 of one]).

Classroom Challenge:

Carol Novillis Larson

University of Arizona, Tucson, Arizona

The Part-Whole Relationship

This challenge activity can be used to assess student's fraction understanding of the part-whole relationship. Tell students that each of the eight pairs of rectangles shown represents what is left of two cakes, one with orange frosting (the lighter shading) and one with green frosting (the darker shading). Ask them, for each pair, to look at what is left of the two cakes and tell which portion is more or whether they are the same. (The cake pans for pairs 1, 3, 4, 5, 7, and 8 are the same size.

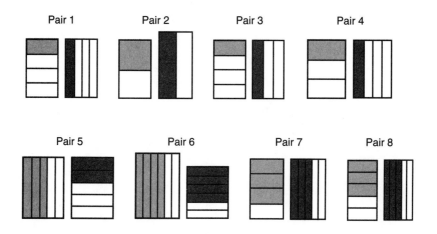

When students were asked to compare these areas, researchers found two major strategies used. In the first strategy, the cakes were cut out, so that student could assure themselves that the cake pans for pairs 1, 3, 4, 5, 7, and 8 were the same size. When students used a *direct comparison* strategy, they would often place the shaded portion of one rectangle on top of the shaded portion of the other and try to guess which shaded area was more. They would say things such as (for pair 7) "Same, because when I put this one over here [student overlaps light-shaded area over dark-shaded area] it was like—I took the extra part from here [light shaded area] and put it with [the student measured with fingers] here. And it's about the same." Note that she compared widths of leftover shaded areas, but ignored differences in lengths.

In the second strategy students compared parts of wholes, thus showing a better understanding of the part-whole relationship. One student, comparing pair 3, said, "Green [dark-shaded] because, like in fractions, one-third would be bigger than one-fourth." For pair 4 another said, "Orange [light-shaded] because this [dark-shaded] is divided up into four slices and this [light-shaded] is only divided into three, and you get a bigger slice because this one is divided up into three pieces."

From: Armstrong, Barbara E. and Carol Novillis Larson. "Students' Use of Part-Whole and Direct-Comparison Strategies for Comparing Partitioned Rectangles". *Journal for Research in Mathematics Education* 26 (January 1995): 2–19.

11

Middle School Introduction

Teresa Yazujian

for the Editorial Panel

T<small>AKE</small> a moment to visualize a preschool classroom. Do you see the activity, hear the noise? Now move to the early elementary grades. Do you see the tubs of materials and students working with all kinds of objects? Can you hear them chattering about what they are doing? Late-elementary school classrooms are still active and noisy as ideas are shared and demonstrated.

Now move in your mind's eye to middle school. Did the noise level and the level of activity change? Is your idea of a middle school level mathematics class different from that of elementary school classes? Is it quieter, less active? I hope you are shaking your head with a resounding NO!

Unfortunately, many middle school classrooms are much quieter and less active than elementary school classrooms. Why? Is it possible that middle school teachers see their role as teacher in a different light from that of their elementary school counterparts?

As you read through the articles in this section, one thesis is continuously repeated. The importance of conversation, the sharing of ideas, is the common thread woven throughout every article. Mathematical ideas are discussed and evaluated through students sharing what they are thinking. Students' misconceptions are brought out as they discuss their ideas with others. How do we learn what students are thinking if we don't ask questions designed to bring out that information?

Prior to beginning work with fractions, I asked my students to tell me everything they knew about fractions. As you would expect, some students had many ideas and others had very few ideas. I consolidated their comments and found that only a small percentage of them was mathematically sound. The following were some of their responses.

- The number in the bottom of a fraction must always be even.
- A fraction is a part of a whole. It can never be a whole. Once it is a whole, it is no longer a fraction.

• The top number is always smaller than the bottom number. If it is not, it is not right.

Though we may wonder how and where students developed ideas such as these, that is not the question on which we need to focus at this point. Our concern at this point is twofold. What activities do we use in the classroom to help students evaluate what they believe to be true mathematically? Also, what activities do we use when developing new concepts to avoid forming additional misconceptions?

The articles in this middle school section provide rich and wonderful discourse. The mathematical concepts being developed as students share their thinking will amaze you. I hope that reading these articles will make you wonder about your own students, their reasoning ability, and what their ideas are with regard to fractions, ratios, and proportions.

Every year I ask my students to "think aloud," and their depth of reasoning never fails to inspire me. Yes, we have to fine-tune some of their conceptual ideas, examining them to find what is sound mathematics and what needs to be reevaluated and changed. It is this reflective thinking, however, that continues to make teaching interesting and may provide a way for our students to change from passive observers to active participants in our classrooms.

12

Percents and Proportion at the Center:
Altering the Teaching Sequence for Rational Number

Joan Moss

CLAIRE and Maggie, two fourth-grade students, were sitting together at a small table with a deck of playing cards. Each girl had half the deck face down in front of her. Their classmates also sat in pairs or in small groups scattered around the classroom. They were all engaged in an assortment of mathematics games that they had invented to practice working with rational numbers in mixed representations. It was the last of a series of eighteen experimental lessons in rational number, and the students were eager to play the games and try the challenges that served as a wrap-up and review.

"One, two, three, go!" Claire and Maggie chorused in unison while each turned up the top card on her pile. The game that they were playing was modeled on the popular card game of war. Both girls stared intently at these two face-up cards. Maggie's card had ".20" written on it and Clare's had "1/8."

"OK," said Claire, "now we have to figure out who has more." Without hesitation, Maggie replied, "I do, 'cause *you* only have 12 1/2 percent but *I* have 20 percent. So *mine* is more."

"Yeah, you're right," Claire agreed. "OK, I have to write down your score.... Hmm, so that's 20 percent take away 12 1/2 percent, so that's 7 1/2 (percent)." Claire then took a pencil, and finding Maggie's place in the score column, she wrote .075. Satisfied, the girls then settled in for the next round of the game.

A visiting teacher in the classroom who had witnessed the interaction from a distance moved over to the two students and asked if they would talk to her a bit about the reasoning that they had used.

Teacher: I was interested to know how you figured out which of the numbers was more, 0.20 or 1/8. First of all, how did you know that 1/8 is equal to 12 1/2 percent?

Maggie: OK, it is like this. One-eighth is half of 1/4, and 1/4 is 25 percent. So, half of that is 12 1/2 percent.

Teacher: Well, you certainly know percents very well. But what about decimals? Do you know what 12 1/2 percent is as a decimal?

Claire: You see, 12 1/2 percent is like point 12 and a half, and that's the same as point 12, point 5, because the point 5 is like half.

Maggie: Yeah, but in decimals you have to say it's really point 125.

What may be immediately apparent from the vignette above is the ease and flexibility that Claire and Maggie had in moving from fractions to percents and then to decimals and back again. As this scenario reveals, Claire and Maggie were able to find the equivalencies for rational number that have proven to be difficult even for much older students (Sowder 1995); they knew that 1/8 = 12 1/2% = .125 and also, as shown when Claire recorded Maggie's score, that 7 1/2 percent can be represented as .075.

A further perusal of the exchanges above also reveals a number of other features of these two students' reasoning—features that were also common to all the other students in this mixed-ability fourth-grade class by the end of the intervention. The first is the use of percents as an intermediate step even when the problem does not contain the percent representation. (In trying to compare 1/8 and 0.20, the girls first converted these quantities to percents; 1/8 = 12 1/2% and 0.20 = 20%.) A second feature, another bridging step, was the use of a double decimal representation. As Maggie explains to the visiting teacher when asked how to convert 12 1/2 percent to a decimal, "…twelve and a half percent is equal to point 12 point 5, and that in turn is equal to point 125." Finally, a third feature of these students' reasoning is the operation of halving and doubling. In order to find the decimal equivalent for 1/8, Maggie first doubled 1/8 to produce the familiar (easier) fraction 1/4 and its percent counterpart of 25 percent. Then she again halved that quantity to get the desired amount of 12 1/2 percent.

How did this kind of reasoning develop? Rational number is often a difficult topic to learn. Students have difficulty in abandoning their whole-number understandings, which lead to many misconceptions in their reasoning (e.g., Carpenter, Fennema, and Romberg 1993). Thus, for example, a typical response to the question "What is 1/8 as a decimal?" would likely be that it is ".8"—because there is an 8 in 1/8 (Moss 2000). However, even when misconceptions that are based on the interference of whole-number thinking are eventually overcome, students' general understanding often remains limited.

Although most students learn to operate with the individual representations of rational number (fractions, decimals, and percents), they fail to make sense of the connections among them (Markovits and Sowder 1994). Clearly the fourth-grade students in this classroom operated in a very different way than many students who are traditionally trained. What might account for these differences?

THE CONTEXT

The twenty-one students in Maggie and Claire's class had all been participants in a research project for the development of rational-number understanding that I designed in collaboration with Robbie Case (Moss and Case 1999). Our hypothesis was that students' everyday knowledge of percents and their intuitions for proportions would serve as a powerful introduction to rational number and would foster a kind of competence with this number system that could be characterized as *number sense* (NCTM 1989, 2000) for the rational numbers. Thus, in our curriculum, we altered the traditional sequence of instruction by beginning with percents. The study of decimals and fractions came later and was grounded in the students' understanding of percents. In this article, I will present the rationale for the curriculum design as well as details of the instructional program that was developed. In addition, I will also illustrate the kinds of reasoning that developed as the students participated. Finally, I will offer further illustrations of the students' reasoning in the context of posttest interviews that were conducted.

VISUAL PROPORTION AND HALVING AND DOUBLING

The curriculum that we developed had its theoretical grounding in Case's theory of intellectual development. Based on that theory, two separate schemas were identified as the basis for this developmentally sequenced curriculum. The first was students' informal, nonnumerical understandings of proportionality, and the other was a numerical flexibility that students of this age acquire for halving and doubling numbers.

Let us first consider the schema for visual proportional evaluation. Although we know that formal proportional reasoning is slow to develop, it has nonetheless been shown that children from a very early age have a strong propensity for making proportional evaluations that are nonnumerical and based on perceptual cues. For example, young children have little difficulty perceiving narrow, upright containers in proportional terms. Although they can see which of two such containers has more liquid in it in absolute terms, they can also see which has more in proportional terms. That is to say, they can see which one is "more full."

What about halving and doubling? As Confrey (1994) points out, halving and doubling have their roots in a primitive scheme that she calls *splitting*. Splitting, she asserts, is based on actions that are purely multiplicative in nature and that are separate from those of additive structures and counting. Whereas in counting the actions are joining, annexing, and removing, in splitting the primitive action is creating simultaneous multiple versions of an original through dividing symmetrically, growing, magnifying, and folding (Confrey 1994, p. 292).

Our proposal, then, is that it is the merging of these two separate multiplicative intuitions—visual proportional evaluation and halving and doubling that forms a core conceptual understanding for rational number. This core understanding first facilitates initial understandings of the individual representations of rational number and eventually allows students to understand how these different representations are related to each other. With these understandings in place, students learn to move among percents, decimals, and fractions freely and fluently, depending on their purpose. More than anything else, it is this flexible movement that demonstrates that children have acquired true number sense and not just a set of isolated conceptual understandings and algorithms.

Implications for Instruction

Following from the preceeding analysis, our task was then to find a context and to design an instructional sequence that would, first, highlight students' intuitive understanding of proportion and their numerical procedures for splitting numbers and, then, foster their interlinkage. The context that we chose was the idea of *fullness* or *amounts relative to the whole*. To quantify these proportional judgments we introduced percents as the first rational-number representation. In the section below, I next present the sequence of activities that took place in a mixed-ability, fourth-grade class.

THE CURRICULUM

Percent in Everyday Life

Since students in this fourth-grade class had not received any formal instruction in percent, we started the experimental lessons with discussions that probed their everyday knowledge of this topic. Not only were the students able to volunteer a number of different contexts in which percents appeared (e.g., their siblings' school marks, price reductions in stores, and taxes on restaurant bills); but their responses also indicated a good qualitative understanding of what different numerical values "meant." For example, students stated that 100 percent meant "everything," 99 percent meant "almost everything," 50 percent meant "exactly half," and 1 percent meant "almost nothing."

Beakers of Water: A Representation for Fullness

We next presented the students with exercises in which they used percent terminology to estimate fullness of cylindrical beakers filled with different amounts of water. ("Approximately what percent of this beaker do you think is full?" or "How high will the liquid rise when it is 25 percent full?"). As it turned out, children's natural tendency, when confronted with the fullness problems, was to use a halving strategy. That is, they determined where a line representing 50 percent would go on the cylinder, then 25 percent, then 12.5 percent, and so on.

The visual estimation exercises using vials and beakers were continued with a new focus on computation and measurement. The children were instructed to compare visual estimates with quantities that they calculated on the basis of measurement and computation. For example, if it were discovered on measuring a beaker that it was 80 mm tall, then 40 mm from the bottom would be the 50 percent point, and 20 mm would be the 25 percent point.

Invented Algorithms

The children were not given any standard rules for performing these calculations, and thus they employed a series of strategies of their own invention. For example, to calculate 75 percent of the length of an 80 cm desktop, the students typically considered this task in a series of steps: *step 1*, find half, and then build up as necessary (50% of 80 = 40); *step 2*, find the difference between 75 percent and 50 percent (75% – 50% = 25%); *step 3*, find 25 percent of 80 (25% × 80 = 20); and *step 4*, sum the parts (40 + 20 = 60). To facilitate this process, we initially presented problems that could be solved precisely by using this general strategy.

Percents on Numberlines

In addition to the beakers, we also included activities with laminated meter-long number lines calibrated in centimeters to provide students with another way of visualizing percent. For example, we incorporated exercises in which children went on "Percent Walks." Here the number lines, which came to be known as "sidewalks," were lined up end-to-end on the classroom floor with small gaps between each one. Students challenged each other to walk a given distance (e.g., "Can you please walk 70 percent of the first sidewalk? Now, how about three whole sidewalks and 65 percent of the fourth?"). The number line activities were used to consolidate percent understandings and to extend the linear measurement context.

Introduction to Decimals

A firm conceptual grounding of decimal numbers is difficult for most students to achieve (Hiebert, Wearne, and Taber 1991). The similarities between the symbol systems for decimals and whole numbers have led to a number of misconceptions and error types. Grasping the proportional nature of decimals is particularly challenging. In our program, we introduced two-place decimals in their relation to percents, again in the context of measurement. The basic approach was to show that a two-place decimal number represents a percentage of the way between two adjacent whole numbers. Or, in linear-measurement terms, a decimal represents an intermediate *distance* between two numbers (e.g., 5.25 is a distance that is 25 percent of the way between 5 and 6).

Decimals and Stopwatches

To begin the lessons in decimals, the students were given LCD stopwatches with screens that display seconds and hundredths of seconds. (The latter are indicated by two small digits to the right of the numbers.) The students were asked to consider what the two "small numbers" might mean and how these small numbers related to the bigger numbers to the left (seconds). After experimenting with the stopwatches, the children noted that there were 100 of these small units of time in one second. With this observation, they made the connection to percents (e.g., "It's like they are percents of a second"). The students came to refer to these hundredths of seconds as "centiseconds," a quantity that they understood as the percentage of time that had passed between any two whole seconds. What we discovered in working with the children was that although the beakers and number lines reinforced the notion of linear measurement, the stopwatches served as a temporal analog of distance.

Magnitude and Order in Decimal Numbers

To illuminate the conceptually difficult concepts of magnitude and order, we devised many activities to help the students actively manipulate the decimal numbers. The first challenge that we presented was the "Stop/Start Challenge." In this exercise, students attempted to start and stop the watch as quickly as possible, several times in succession. They were taught to record their times as decimals. So, for example, 20 centiseconds were written as .20, 9 centiseconds as .09, and so on.

Next, they compared their personal quickest reaction time with those of their classmates, and they had the opportunity to experience the ordering of decimal numbers as well as to have an informal look at computing differences in decimal numbers (scores). Another stopwatch game that offered active participation in the understanding of magnitude was "Stop the Watch Between" (e.g., "Can you stop the watch between .45 and .50?"). Finally, in another game, called "Crack the Code," students moved between representations of rational number as they were challenged to stop the watch at the decimal equivalent of different fractions and percents. For example, if given the secret code "3/4," then they were required to stop the watch at 75 centiseconds. The students enjoyed these games and invented many others to challenge their classmates.

Fractions and Mixed Representations

A special word must be said about fractions. Fraction terminology was used throughout the program, but only in relation to percents and decimals. At the beginning, all of the children naturally used the term *one-half* interchangeably with *fifty percent*, and most knew that 25 percent (the next split) could be expressed as one-quarter. We also told them that the 12 1/2 percent split was called *one-eighth* and showed them the fraction symbol 1/8. In the final lessons of the experimental instructional sequence, we included fractions in a series of activities that were designed to give the students experience in working with multiple representations of rational number. The students identified equivalencies, ordered quantities in mixed representations, and performed simple computations with these numbers. These activities included—

1. true or false exercises such as "0.375 is equal to 3/8, true or false?"
2. stopwatch games with instructions like "stop the watch [the LCD stopwatches described above] as close to the sum of (1/2 + 3/4) as possible, (i.e. 1.25), and then figure out the decimal value for how close you were."
3. challenge addition, in which students were asked to invent a long, mixed-addition problem, such as 1/4 + 25% + 0.0625 + 1/16, with which to challenge their classmates.

4. specially designed playing cards students used to explore mixed representation further as illustrated in the opening vignette. A complete list of both the content covered in each lesson and the challenge problems that were assigned is presented in Moss (2000).

CHILDREN'S ACQUIRED UNDERSTANDINGS

As can be seen in the description of the curriculum above, the students adapted naturally to the exercises that were presented. From the very first lessons, they showed an everyday knowledge of percents that was underpinned by intuitions for proportion. Further, as the students engaged in exercises using the beakers of water, we also discovered that they had considerable informal knowledge of other concepts that are central to rational-number understanding; for example, students successfully incorporated ideas of part/whole, and equivalencies, and also appeared to have a grasp of the referent unit and its transformations. Similarly, when decimals were introduced in the contexts of stopwatches, the students could make sense of this new representation and were able to perform a variety of computations. Finally, as was illustrated in the opening vignette, by the end of the eighteen experimental sessions, the students had gained a flexible approach to translating among the representations of rational number, using familiar benchmarks and halving and doubling as vehicles of this movement. In fact, what we repeatedly observed as the curriculum unfolded was that when the students worked in the particular contexts that we provided and worked with numbers that lent themselves to operations of halving and doubling, they were successful. What we were *not* able to assess in the course of the lessons, however, was how students would perform on tasks that were different from those presented in the curriculum. On the one hand, we were interested in finding out how students would perform on tasks that used unfamiliar contexts, and on the other hand, we wanted to discover how the students would perform on tasks that incorporated numbers that were more difficult for students to manipulate.

Posttest Interviews

In the final section of this article, I offer two illustrations of student's reasoning on tasks that were designed to probe students' ability to work in novel contexts. The two items that I will present were taken from posttest interviews that we conducted as part of the research project. In the first item that I present, students were required to choose which was the larger fraction, 1/2 or 1/3, and then to find a number that might come between these two fractions. We were interested in discovering if students could work with the fraction 1/3—a fraction that was not featured in their classroom instruction—as well as in probing students' understanding of the density property of rational number. To illustrate how students responded to this challenge, I present the reasoning of two students whose responses are representative of most of the students in the class. Katie, the first of the respondents, was a lower-achieving student, and the other student, Andy, had been identified as a high achiever.

Interviewer: Which do you think is more, 1/2 or 1/3?

    Katie: One-half.

Interviewer: Why is that?

    Katie: Well, if you drink a half a glass of juice, you get more than if you drink a third of a glass, 'cause halves are always more. Well, 'cause you divide the cup in two for halves and three for thirds.

Interviewer: Now, can you tell me if you think that another fraction could fit between 1/2 and 1/3?

    Katie: I know that there are fractions, but I am not sure what they are. Maybe 3/8. I am not sure, but I think it is right. Can I use the paper? [*Here Katie held a standard-sized sheet of paper horizontally and then folded it into eight long equivalent strips.*] Well, this [*pointing to three adjacent eighths*] is 3/8. [*Next, she took another piece of paper, and holding it in the same way, she carefully folded it into thirds. She then aligned the two pages and compared the 3/8 to the 1/3.*] You see 3/8 is a bit bigger.

Andy responded to these same questions in a different way.

    Andy: One-half is bigger because 1/2 is equal to 50 percent and 1/3 is 33.3 percent …. Yes, there are many numbers that could fit between [1/2 and 1/3] like 38 percent or something. So, as a fraction, that would be 38/100.

As can be seen, both of the students were able to provide reasonable answers to the two questions that had been posed. Although Katie used concrete examples of quantity to reason about the two questions, what was apparent in her responses was the strong multiplicative basis in her reasoning. Andy, on the other hand, reasoned in ways that were more representative of the curriculum and used equivalencies to determine his answers.

The second posttest item that I will discuss asked students to compute mentally a percent of a given quantity, namely, 65 percent of 160. Although this type of computation was regularly performed in our classroom, the numbers that were involved were significantly more difficult than those the students typically encountered in their lessons. Furthermore as can be seen in the protocol below, this item required that students work with 10 percent and with the familiar benchmarks (25 percent, 50 percent, 75 percent, and 12 1/2 percent) that served as a basis for most of their classroom work. As the two examples below reveal, the students were able to find ways to solve this difficult problem.

Interviewer: What is 65 percent of 160?

    Sascha: Okay, 50 percent of 160 is 80. Half of 80 is 40 so that is 25 percent. So if you add 80 and 40 you get 120. But that [120] is too much because that's 75 percent. So you need to minus 10 percent [of 160] and that's 16. So, *120 take away 16 is 104.*

    Neelam: The answer is 104. First I did 50, percent which was 80. Then I did 10 percent of 160, which is 16. Then I did 5 percent, which was 8. I added them [16 + 8] to get 24, and added that to 80 to get 104.

CONCLUSION

Rational Number Sense

In this research, we designed a curriculum that featured percent in linear measurement as an introduction. The curriculum was based on the hypothesis that a flexible understanding of the rational number system was underpinned by a core conceptual structure that is formed by the merging of students' intuitions for visual proportional assessment and their schemas for halving and doubling. As can be seen in the posttest procedure as well as in the description of the reasoning developed in the course of the instructional unit, students gained a flexible approach to this number system that can be characterized as number sense. The scope of students' acquired understandings included—

1. the ability to use the representations of decimals, fractions, and percents interchangeably;
2. an appreciation of the magnitude of the rational numbers as seen in students' ability to compare and order numbers within this system;
3. the ability to invent a variety of solution strategies for calculating with these numbers;
4. a general confidence and fluency in student's ability to think about the domain, using the benchmark values that they have learned.

Percent as a Starting Point for Instruction

One of the implications of this work, then, is the usefulness of percent and its privileged base of 100 in early rational-number learning. Grade 4 students' extensive knowledge of the numbers from 1–100 naturally facilitates comparison questions as well as promotes students' ability to translate among the representations of rational number—any percent value can be translated into a fraction or a decimal. The converse, however, is not easily done. The ability to invent strategies for calculating is also facilitated by the percent construct. A survey conducted by Lembke and Reys (1994) revealed that students who had not yet had formal training in percents were able to invent procedures to solve percent tasks and were even better able to solve certain kinds of operations with percents than were older students who had learned percent in school.

To my knowledge, no other curriculum uses percent as an introduction. In fact percent is not introduced until well after students have learned decimals and fractions, and this can be as late as the sixth grade—and even then many students have a great deal of difficulty (Parker and Leinhardt 1995). The results of the present study show a very different learning pattern with regard to percents. Not only did the students' ability to perform percent tasks improve with instruction—but it was also discovered that students had substantial intuition about percent meanings and operations prior to instruction. Finally, as we continued our work with these fourth-grade students, we also discovered that percents provided a solid multiplicatively based foundation for the subsequent learning of decimals and fractions and thus served as a useful starting point for an overall understanding and flexibility of the rational-number system as a whole.

REFERENCES

Carpenter, Thomas P., Elizabeth Fennema, and Thomas A. Romberg. *Rational Numbers: An Integration of Research.* Hillsdale, N.J.: Lawrence Erlbaum Associates, 1993.

Confrey, Jere. "Splitting, Similarity, and the Rate of Change: A New Approach to Multiplication and Exponential Functions." In *The Development of Multiplicative Reasoning in the Learning of Mathematics,* edited by Guershon Harel and Jere Confrey, pp. 293–332. Albany, N.Y.: State University of New York Press, 1994.

Hiebert, James, Diane Wearne, and Susan Taber. "Fourth Graders' Gradual Construction of Decimal Fractions during Instruction Using Different Physical Representations." *Elementary School Journal* 91 (1991): 321–41.

Lembke, Linda, and Barbara J. Reys. "The Development of, and Interaction between, Intuitive and School-Taught Ideas about Percent." *Journal for Research in Mathematics Education* 25 (May 1994): 237–59.

Markovits, Zvia, and Judith Sowder. "Developing Number Sense: An Intervention Study in Grade 7." *Journal for Research in Mathematics Education* 25 (January 1994): 4–29.

Moss, Joan. "Deepening Children's Understanding of Rational Number: A Developmental Model and Two Experimental Studies." Doctoral diss., University of Toronto, 2000.

Moss, Joan, and Robbie Case. "Developing Children's Understanding of the Rational Numbers: A New Model and an Experimental Curriculum." *Journal for Research in Mathematics Education* 30 (March 1999): 122–47.

National Council of Teachers of Mathematics (NCTM). *Curriculum and Evaluation Standards for School Mathematics.* Reston, Va.: NCTM, 1989.

———. *Principles and Standards for School Mathematics.* Reston, Va.: NCTM, 2000.

Parker, Melanie, and Gaea Leinhardt. "Percent: A Privileged Proportion." *Review of Educational Research* 65, no. 4 (1995): 421–81.

Sowder, Judith. "Instructing for Rational Number Sense." In *Providing a Foundation for Teaching Mathematics in the Middle Grades,* edited by Judith Sowder and Bonnie Schappelle, pp. 15–29. Albany, N.Y.: State University of New York Press, 1995.

13

Making Explicit What Students Know about Representing Fractions

Barbara M. Moskal

Maria E. Magone

ONE purpose of this article is to illustrate how an open-ended task can be used to elicit evidence of the appropriate and inappropriate relationships that students have established within the fractional number system. "Open ended" refers to a task structure that allows students to determine their own approach when solving problems. Open-ended tasks typically include a request for a display of either the student's approach or reasoning process. An advantage of open-ended tasks is that they can offer detailed information concerning students' knowledge.

Eight open-ended tasks that were developed as part of the Quantitative Understanding: Amplifying Student Achievement and Reasoning (QUASAR) project's Cognitive Assessment Instrument (QCAI) (Lane et al. 1995) were administered to fifth-grade students (Moskal 1997). QUASAR was a mathematics reform project at the University of Pittsburgh designed to improve the mathematical knowledge of middle school students in disadvantaged communities. Of these eight tasks, one—the Fractional Representation Task, or "Pizza"—was designed to assess students' understanding of fractions. Examples of students' responses to this task will be presented and discussed in order to illustrate the type of detailed information that an open-ended task can provide with respect to a student's knowledge of fractions.

The influence of both students' prior experiences and the results of research should be considered when interpreting students' responses to open-ended tasks. In our discussion of these responses, careful consideration has been given to the impact that students' prior knowledge and out-of-school experiences have on their interpretations of the problems and their responses to them. Careful thought has also been given to the research that describes students' development of fractional knowledge.

ANOSH'S RESPONSE

Anosh provided a correct numerical answer for both of the questions in the task (see fig. 13.1). Anosh separated the large pizza into 12 pieces and determined that Elena would get 1/3 of the pizza. According to this answer, Anosh apparently recognized that Elean would receive 4/12 of the pizza. He then either reduced the 4/12 to 1/3 or recognized that 4 pieces of pizza composed 1/3 of the whole pizza.

In the second question of this task, Anosh's response was, "1 hole & 3 pecies" [*sic*]. Anosh's diagram of the segmented pizza suggests that 1 whole pizza and 3 slices would result in 1 1/3 pizzas, a correct answer. Since the task did not specifically require a fractional representation, Anosh's answer is acceptable. From Anosh's response, it cannot be determined whether he understands how to express fractions that are larger than the unit. This raises an important concern when using open-ended tasks: The manner in which the question is posed and the manner in which the student selects to respond may limit the information that is acquired in the analysis process.

Luis, Elena, and Leslie plan to share 1 large, square pizza. Each person will get an equal amount.

A. Show on the picture how much pizza Elena will get.

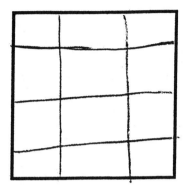

B. What fraction of the pizza will Elena get? $\frac{1}{3}$

Answer: $\dfrac{1}{3}$

Maria, Carlos, and Terry wanted to share 4 medium, square pizzas. Each person will get an equal amount.

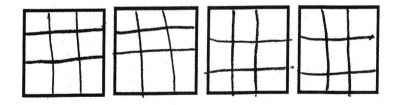

A. Show on the picture how much pizza Carlos will get.
B. How many pizzas will Carlos get?

Answer: 1 hole + 3 pecies

Fig. 13.1. Anosh's response

JANEL'S RESPONSE

Janel's response is shown in figure 13.2. For the first question, Janel divided the large pizza into 3 slices and correctly indicated that Elena would receive 1/3 of the pizza. For the second question, Janel divided each of the four medium pizzas into four parts, shaded one part of each pizza, and indicated that Carlos would receive 4/4, or 1, pizza. A positive feature of Janel's response is that she recognizes that 4/4 is equal to 1. However, if each of the three students received one pizza, there would be an entire pizza left over.

There are several possible explanations for Janel's response. The statement of the task does not explicitly say that the 4 pizzas should be completely divided among the three students. In this interpretation, Janel's answer reflects a correct alternative solution to the given task. Each student would receive one pizza, and one pizza would be left over. Another interpretation is that Janel may not have understood how to express a fraction that is greater than 1. This is a common difficulty for students as they begin to learn fractions. Janel may have eliminated the need to express a fraction that was greater than 1 by restricting each person to a maximum of one pizza. Another possibility is that Janel may have restricted herself to the types of subdivisions that she had seen in her textbook (e.g., 1/2, 1/4, 3/4, 1/8). When the pizzas were subdivided into four parts, there was no simple solution to the process of evenly distributing the pizza to the individuals represented in the task.

Janel's response may be the result of a limited understanding of the concepts that are being assessed or of an alternative interpretation of the task constraints. In order to determine the appropriate interpretation of Janel's response, her teacher may decide to probe her understanding verbally. This illustrates an important benefit of classroom assessment. When a student's response does not clearly indicate the nature of the student's knowledge, the teacher may use follow-up assessment activities, such as verbal probing, to better understand the student's knowledge structure.

Luis, Elena, and Leslie plan to share 1 large, square pizza. Each person will get an equal amount.

A. Show on the picture how much pizza Elena will get.

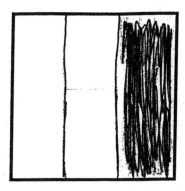

B. What fraction of the pizza will Elena get?

Answer: $\dfrac{1}{3}$

Maria, Carlos, and Terry wanted to share 4 medium, square pizzas. Each person will get an equal amount.

A. Show on the picture how much pizza Carlos will get.
B. How many pizzas will Carlos get?

Answer: $\dfrac{4}{4} = 1$

Fig. 13.2. Janel's response

ERIC'S RESPONSE

Eric's response is shown in figure 13.3. He responded incorrectly to both the questions. An examination of his responses suggests that his difficulties may have resulted from the combined impact of his prior experiences and his inadequate written interpretation skills.

Eric sought to divide the pizza into triangular segments that are similar to how a circular pizza is cut. This suggests that Eric was trying to make sense of the task using his prior experiences. In part B of the first question, Eric's answer of 1 1/3 reflects the fraction of slices (assuming that the large triangular slice is the unit) that Elena would receive rather than the fraction of the pizza. In other words, Eric may have redefined the unit to be a large triangular slice. He did not realize that in the statement of the task, "What fraction of the pizza will Elena get?" the unit was already defined to be a whole pizza.

For the second question, Eric once again ignored the criterion that the unit of measure was a whole pizza. In his response, Eric divided three of the pizzas in half and the third pizza into six unequal wedges. His answer was "4 pices" [*sic*]. It is difficult to determine from Eric's work whether he did not understand the statement of the task or whether he did not understand the concepts being assessed in the task. His response in part A does indicate some understanding of fractional representations.

Luis, Elena, and Leslie plan to share 1 large, square pizza. Each person will get an equal amount.

A. Show on the picture how much pizza Elena will get.

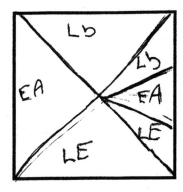

B. What fraction of the pizza will Elena get?

Answer: 1⅓

Maria, Carlos, and Terry wanted to share 4 medium, square pizzas. Each person will get an equal amount.

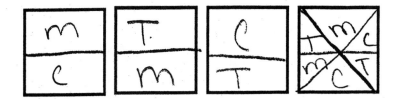

A. Show on the picture how much pizza Carlos will get.
B. How many pizzas will Carlos get?

Answer: 4 pices

Fig. 13.3. Eric's responses

Conclusion

Classroom assessments should support two goals (Van den Heuvel-Pan-huizen 1994). First, assessments should furnish information that allows teachers to look back on the effectiveness of the instruction that they have presented. On the basis of the responses that were examined in this article, a teacher could determine the extent to which his or her instruction was effective in providing students with a basic understanding of how to apply fractional knowledge to the given situation and how to represent the resultant fractional values. If we assume that these outcomes were consistent with the instructional goals, the students' responses suggest the extent to which these goals were reached. Second, assessment should offer information that is useful for guiding future instruction. The students' responses to the given tasks indicate what the students knew and did not know with respect to fractions. Their teacher may use this information to guide future instruction that builds on students' correct conceptions and challenges their misconceptions.

Teachers must also be aware of the limitations of open-ended tasks. As was illustrated by Anosh's response, students may interpret a task in a manner different from what was intended. Although Anosh's response is appropriate under his alternative interpretation, it fails to provide the teacher with desired information.

Another difficulty is when a student gives a response that can be interpreted in multiple ways. In Janel's response, it was unclear whether she did not understand the mathematics involved in the task or whether she was confused by the implied constraints of the problem. A teacher working in isolation may not recognize the different possible interpretations of Janel's response. To increase the likelihood that alternative interpretations are identified, a teacher may find it useful to discuss some students' responses with another teacher. A comparison of both interpretations will result in greater insight into the student's thinking.

Open-ended tasks are often embedded within a context. This, too, can cause difficulties. Problems that are embedded in a context have the benefit of illustrating to students the practical applications of mathematics. Also, contextualized problems allow students to use their prior experiences in the solution process. However, the specific context may cause students to impose constraints on the problem that were not intended. For example, Eric initially attempted to divide the pizza into triangular pieces. This constraint resulted in a more difficult problem. Another example is shown by Anosh's efforts to divide the initial pizza into twelve slices rather than three slices. Real-life pizzas are often divided into eight or twelve pieces, and Anosh may have been using his past experience with pizzas to create pizza slices that were small enough to eat. This is a constraint that was not intended in the task.

In all the examples above, a follow-up assessment activity could be used to clarify the students' thinking. For example, a teacher could ask Janel to explain her reasoning verbally. Another possibility is that the teacher could examine Janel's responses to multiple assessment tasks. This approach would allow her teacher to identify patterns that emerge in Janel's reasoning across different tasks and thus eliminate some of the guesswork that was involved in interpreting an isolated response. This approach raises an important point with respect to classroom assessment: Classroom assessment should be an ongoing activity in which the results of prior assessments guide both the selection and interpretation of follow-up assessment activities.

In this article, responses from three different students were examined in great detail. Both the students' prior experiences and the results of research were used to guide the interpretation process. It is unreasonable to expect teachers to examine a classroom set of responses in as much depth. However, a teacher could examine in detail a random sample of seven or eight responses collected from a classroom of students and discuss these sample responses with another teacher. The analysis of these responses is likely to suggest both the types of errors to look for across students' responses and the limitations that are inherent in a given task. Teachers could then use this information to direct their attention in a focused analysis of the remaining responses. The end result would be greater insight into the understandings and misunderstandings of students' fractional knowledge that prevail within the classroom. This information could then be used to guide future instruction.

REFERENCES

Lane, Suzanne, Edward A. Silver, Robert D. Ankenmann, Jinfa Cai, Connie Finseth, Mai Liu, Maria E. Magone, David Meel, Barbara M. Moskal, Carol S. Parke, Clem A. Stone, Ning Wang, and Yuehua Zhu. *QUASAR Cognitive Assessment Instrument (QCAI)*. Pittsburgh, Pa.: University of Pittsburgh, Learning Research and Development Center, 1995.

Moskal, Barbara M. "Written Open-Ended Decimal Tasks: What Information Do Teachers Acquire and Use through the Examination of the Modes, Referents and Relationships that Students Employ?" Ed. D. diss., University of Pittsburgh, 1997.

Van den Heuvel-Panhuizen, Marja. "Improvement of (Didactical) Assessment by Improvement of Problems: An Attempt with Respect to Percentage." *Educational Studies in Mathematics* 27 (December 1994): 341–72.

14

Using Literature as a Vehicle to Explore Proportional Reasoning

Denisse R. Thompson

Richard A. Austin

Charlene E. Beckmann

As INDICATED in the *Principles and Standards for School Mathematics* (National Council of Teachers of Mathematics 2000), the development of proportional reasoning is a major focus of the middle-grades curriculum. Although much of the formal study of proportions occurs in the middle grades, much informal work with proportional reasoning through multiplicative situations is appropriate in the upper elementary grades. The challenge for educators is to find contexts that engage students and that facilitate the study of proportional reasoning, either formally or informally.

Literature provides a natural context in which this study can occur. A number of authors have focused on the use of literature at the elementary school level to engage students in exploring mathematics (Whitin and Wilde 1992, 1995). Some work is beginning to appear about the use of literature to engage middle grades students in studying mathematics (Austin and Thompson 1997; Lawrence 1999).

A number of literature titles exist in which the underlying story centers on proportional thinking. At the upper elementary grades, these books can be used to study ratios and proportions informally by focusing on the multiplicative relationships inherent in the story. At the middle grades, these same books can be used to study ratios and proportions more formally. We have used a number of these books to explore proportional thinking with students, informally in grades 3–5 and both formally and informally in grades 6–8.

MULTIPLICATIVE REASONING IN THE ELEMENTARY GRADES

With students in the intermediate grades, we have focused on proportional reasoning not through formal study of ratios and proportions but informally through multiplicative ideas. Consider the book *Jim and the Beanstalk* (Briggs 1970). In this story, Jim is the son of Jack (from Jack and the Beanstalk fame) and the giant is the son of the giant from the earlier story. But the current giant has problems: he cannot read because of eye difficulties, he cannot eat because of teeth problems, and he believes he is ugly because he has no hair. To solve each problem, Jim takes appropriate measurements and has objects made to solve the giant's problems. For instance, to solve the eye problem, Jim measures the giant's head, and with a large coin pays to have a pair of reading glasses made for the giant. In similar fashion, Jim has dentures and a wig made to solve the other problems.

We have used this book with fifth-grade children to consider proportional reasoning by focusing on multiplicative relationships. Suppose the giant is 9 feet (108 inches) tall, and that one of the children, Lynne, is 51 inches tall. We have asked the children a question similar to the following: "The giant is how many times as tall as Lynne?" That is, "How many Lynnes would need to be stacked on top of each other to equal the height of the giant?"

Children have approached this problem in several ways. Initially, some consider additive strategies, specifically adding Lynne's height until they reach the giant's height, or come fairly close to the giant's height. Others quickly consider multiplicative strategies by multiplying Lynne's height by successive numbers until they obtain a value as close as possible to the giant's height. In this case, they realize that $51 \times 2 = 102$ inches is too small but that $51 \times 3 = 153$ inches is too large. Successive iterations with multipliers between 2 and 3 permit the children to come fairly close to the giant's height. Typically, in considering how to determine the exact number of Lynnes needed to equal the giant's height, some child considers dividing the heights (i.e., $108 \div 51$). Thus, the giant is about 2.12 times Lynne's height.

Helping children at this level come to grips with multiplicative reasoning is not trivial. We have found that many children struggle with the idea of "A is how many times as big as B?" But by translating the question to "How many A's would need to be stacked on top of each other or laid end to end to equal B?", children are able to engage in this process of multiplicative reasoning. Further, in this process of reasoning, with its associated computation, they can also explore other mathematical concepts, such as measurement.

At this stage, we typically have children measure some common object that they regularly use, such as a toothbrush or a comb. They determine their height compared to the length of the toothbrush or the comb—namely, how many toothbrushes or combs would need to be stacked in order to equal their height. Using this value, they determine the appropriate length for the giant's toothbrush or comb in order for him to have the same comparison with his toothbrush that they have with theirs. Although ratios and proportions are not used formally in solving these problems, they are the underlying concepts inherent in the lesson. It is hoped that the informal work done with multiplicative relationships here will provide a foundation on which more-formal work with ratios and proportions can build in the middle grades.

The same multiplicative ideas help children at this level make scale drawings. We have had children locate an object (animal, plant, or building) to which they want to compare themselves. The children draw a picture of themselves 1 inch tall and a picture of the other object in the proper ratio.

For instance, one child chose to compare herself to a sun jellyfish. The book *Biggest, Strongest, Fastest* (Jenkins 1995) states that a sun jellyfish may be 200 feet long as it drags its tentacles through the water. Because the jellyfish is roughly 39 times as tall (long) as she is, a drawing of the jellyfish needs to be 39 inches if she is drawn 1 inch tall. On the classroom floor, she taped multiple pieces of paper together in order to construct her drawing. In a similar manner, other children have compared themselves to large objects.

Throughout these activities, the children were engaged in mathematics beyond what they were studying at that point in their regular mathematics curriculum. Such informal work with multiplicative reasoning has the potential to develop the type of fluency with numbers that we want children to have as they move into the middle grades.

PROPORTIONAL REASONING IN THE MIDDLE GRADES

At the middle grades, we have used a modified version of the Jonathan Swift classic about Gulliver, *Gulliver in Lilliput* (Hodges 1995), to engage middle grades students in activities focusing on proportional thinking. The modified version tells only the tale of Gulliver in the land of the Lilliputians. A number of children have seen all or part of the television miniseries about Gulliver, so that mathematics lessons based on this tale have a somewhat familiar setting. (An Internet site contains an edited version of *Gulliver's Travels* – www.jaffebros.com/lee/gulliver/bk1.)

The Lilliputians, who are only 6 inches tall compared to Gulliver's height of 6 feet (our choice for his height), make clothes for Gulliver by taking just two measurements, specifically the distance around the thumb and the length from his neck to the middle of his leg. The Lilliputians have found that twice around the thumb is the distance around the wrist, twice around the wrist is the distance around the neck, and twice around the neck is the distance around the waist. When asked why the Lilliputians would only want to take a single girth measurement, middle grades students recognize that such small people would have a great deal of difficulty taking the measurements of such a large person, relatively speaking. That is, they seem to recognize that the proportional difference between Gulliver and the Lilliputians necessitates some considerations other than direct measurement.

We have had middle grades students, from grades 6–8, measure each of the lengths (thumb, wrist, neck, and waist) and record the results on the board, recording the measures separately for males and females. Table 14.1 contains data from one class of seventh-grade students. Although most students had no difficulty recording measurements on the board, even waist measurements, teachers should be sensitive to those students who are hesitant to include waist measurements, especially given that adolescents are often quite weight-conscious. Observe that for many of the students, the 2:1 ratios of the Lilliputians are in the ballpark, and for other students, the 2:1 ratio would yield rather inaccurate results. Hence, students can consider whether or not a single measure of their thumb, along with a length measurement, would permit a tailor to make a jacket with a reasonable chance of fitting.

TABLE 14.1.
Measures (in cm) from seventh-grade students to check the Lilliputian rule of thumb

Female				Male			
Thumb	Wrist	Neck	Waist	Thumb	Wrist	Neck	Waist
7.5	15	31	60	7	14.2	28.5	57.2
7.5	16	34	78	7	14	29	70
6.5	15	33	69	7	14	32.5	70
8	14	30	65	8	13	28	57
6	14	34	67	7	15	28	60
7.5	14	30.5	60	7.5	13.8	28	59
7.5	15	30	62.5				

In addition to taking measurements and comparing to the stated 2:1 ratio, students have graphed results, such as thumb-versus-neck measure. Notice that such a graph is roughly linear and can be described by the equation

$$\text{neck distance} = 4 \cdot \text{thumb distance.}$$

Likewise, students can determine the ratios that exist between waist and thumb or between waist and wrist.

As mentioned earlier, in the modified story the Lilliputians are 6 inches tall and Gulliver is 6 feet tall. Hence, the corresponding scale is 1 to 12. We have typically had students make a scale drawing of themselves, using a 1 to 10 or 1 to 20 scale so that their drawing fits on a regular sheet of paper. Although many students determine their scale drawing height by simply dividing their height by 10 or 20, respectively, the underlying mathematics involves proportional reasoning to realize that each 10 centimeters of their height is drawn as only 1 centimeter on paper.

In constructing their scale drawing, many focus only on their height in the scale drawing. Nevertheless, they make a drawing of themselves with reasonable proportions in terms of head, neck, legs, and so on. Other students have actually measured the length of their head, their neck, or their legs to guarantee that their scale drawing shows their body in the same proportion as in real life.

Another book that permits interesting comparisons for middle grades students is *Biggest, Strongest, Fastest* (Jenkins 1995). The book contains interesting facts about animals that set a record in their respective category. For instance, the best jumper for its size is the flea. Although a flea is only 1/16 inch tall, it can jump 8 inches into the air. Students can determine the ratio of the height jumped by the flea to its body height. How high could they jump if their jump height to body height were in the same ratio? If a typical story of a building is 10 feet, students are immensely impressed with themselves to think they could jump 65 to 70 stories.

We have found middle grades students to become quite engaged in doing mathematics as a result of reading one of these books. Not only do they like the stories, but also the contexts seem to offer insight into ways to engage in the mathematics. That is, the contexts make the mathematics less abstract and more concrete.

Literature provides an effective vehicle for engaging in mathematics, not only at the elementary grades but at the middle grades as well. Indeed, the mathematics embedded in many books connects well with the mathematics curriculum of grades 6–8. We encourage teachers to search actively for literature that can be used to form the basis of mathematics lessons or enhance the study of important concepts.

REFERENCES

Austin, Richard A., and Denisse R. Thompson. "Exploring Algebraic Patterns through Literature." *Mathematics Teaching in the Middle School* 2 (February 1997): 274–81.

Briggs, Raymond. *Jim and the Beanstalk.* New York: Putnam & Grosset, 1970.

Hodges, Margaret. *Gulliver in Lilliput.* New York: Holiday House, 1995.

Jenkins, Steve. *Biggest, Strongest, Fastest.* New York: Ticknor & Fields Books, 1995.

Lawrence, Ann. "From *The Giver* to *The Twenty-One Balloons*: Explorations with Probability." *Mathematics Teaching in the Middle School 4* (May 1999): 504–09.

National Council of Teachers of Mathematics (NCTM). *Principles and Standards for School Mathematics.* Reston, Va.: NCTM, 2000.

Whitin, David J., and Sandra Wilde. *Read Any Good Math Lately? Children's Books for Mathematical Learning, K–6.* Portsmouth, N.H.: Heinemann, 1992.

————. *It's the Story That Counts: More Children's Books for Mathematical Learning, K–6.* Portsmouth, N.H.: Heinemann, 1995.

ANNOTATED BIBLIOGRAPHY

Ash, Russell. *Incredible Comparisons.* London: Dorling Kindersley, 1996.

This book provides comparisons for many different categories, such as the surface of the earth, going into space, great lengths, and animal speeds. Students can compare themselves to an appropriate benchmark or compare various objects in a given category. A wide range of ratios and proportions can be constructed from the facts in the book.

Beneduce, Ann Keay. *Gulliver's Adventures in Lilliput.* New York: Putnam & Grosset, 1996.

Gulliver lands in Lilliput, where all the people are only six inches tall. This is a slightly different version of the story from the one mentioned in the narrative of this chapter. Similar activities as those described in the chapter can be explored.

Briggs, Raymond. *Jim and the Beanstalk.* New York: Putnam & Grossett, 1970.

Jim meets the giant, Jack, and takes measurements of his head, his eyes, and his mouth in order to have a wig, a pair of glasses, and a pair of dentures made for the giant. Students can find corresponding measurements for classmates and determine a wide range of ratios similar to those in the chapter.

Carle, Eric. *The Grouchy Ladybug.* New York: Harper Trophy, 1977.

A ladybug flies from animal to animal, always wanting to start a fight. Even though the animals get larger as the ladybug's journey progresses, the ladybug always thinks the animal is too small to fight. Students can determine the number of ladybugs needed to measure a given object, using the length of a ladybug as the basis for the comparisons.

Clement, Rod. *Counting on Frank.* Milwaukee: Gareth Stevens Publishing, 1991.

The boy studies sizes and all types of facts. For instance, he determines how many of his dog would fit in a room, how long it would take to fill a bathtub, or how tall he would be if he grew at a given rate. The various rates mentioned in the book provide a context for exploring proportional thinking.

Hodges, Margaret. *Gulliver in Lilliput.* New York: Holiday House, 1995.

This book is a modified version of the Jonathan Swift classic, *Gulliver's Travels*. Activities similar to those in the chapter can be explored using the information about Gulliver's stay in Lilliput.

Jenkins, Steve. *Biggest, Strongest, Fastest.* New York: Ticknor & Fields Books, 1995.

This book presents information about a number of animals that are record holders in some way. Comparative information is also provided from which a variety of ratios and proportion problems can be developed.

Kellogg, Steven. *Much Bigger than Martin.* New York: Dial Books, 1976.

A young boy always feels too small when compared to his brother. He imagines what it would be like to be considerably bigger than his brother. Further, the book discusses some activities he could do if he were very tall, as well as some activities that he would not be able to pursue. Students could investigate how the sizes of buildings and other objects would need to change to accommodate various large individuals. Students could also investigate world records of "giants" in our lifetime.

Lasky, Kathryn. *The Librarian Who Measured the Earth.* Boston: Little, Brown & Co., 1994.

This is the story of Eratosthenes and how he measured the circumference of the Earth. Students can use ratios to determine distances walked by the bematists and determine how proportions were used to estimate the size of the earth.

Lord, John Vernon. *The Giant Jam Sandwich.* Boston: Houghton Mifflin Co., 1972.

A town makes a giant jam sandwich to capture pesky wasps. Students can explore the amount of ingredients needed to make items that appear in the *Guinness Book of World Records.* Ratios and proportions in expanding or shrinking recipes can also be investigated.

Malam, John. *Highest, Longest, Deepest: A Fold-Out Guide to the World's Record Breakers.* London: Simon & Schuster, 1996.

This book gives many natural examples that are record breakers for being the longest, highest, or deepest in their category. As with other comparison books, students can construct comparisons that are of interest to them, using informal multiplicative reasoning or more-formal proportions.

Most, Bernard. *How Big Were the Dinosaurs?* San Diego: Voyager Books, 1994.

Twenty different dinosaurs are introduced. Their sizes are compared to modern-day objects, providing multiple opportunities for the study of ratios and proportions.

Myller, Rolf. *How Big Is a Foot?* New York: Dell Publishing, 1962.

An apprentice tries to make a bed for the queen. Until he measures with the same foot size as the king, he is not able to build a bed that is the proper size. Not only does the book illustrate the importance of standard versus nonstandard measures, but also the book lends itself to ratio comparisons when different nonstandard measures are used.

Suyeoka, George, Robert B. Goodman, and Robert A. Spicer. *Issunboshi.* Aiea, Hawaii: Island Heritage Publishing, 1974.

A one-inch boy protects a princess and wins her heart and hand. Students can consider appropriate sizes for a boat, a sword, and so on for a person who is only one inch tall. How loud would someone that size need to speak in order for a regular-sized person to hear them?

Wells, Robert E. *Is a Blue Whale the Biggest Thing There Is?* Morton Grove, Ill.: Albert Whitman & Co., 1993.

This book starts with the size of a blue whale and makes comparisons with larger and larger objects as it considers the size of the universe. Relative sizes of the planets, distances between the planets, and travel at the speed of light are all possible topics based on this book. Students can explore the astronomical comparisons in which measures are based on an earth measure of 1.

———. *What's Faster than a Speeding Cheetah?* Morton Grove, Ill.: Albert Whitman & Co., 1997.

Speeds of different objects are compared, including animals, planes, spaceships, and so on. The various facts lend themselves to a variety of ratio and proportion comparisons.

———. *What's Smaller than a Pygmy Shrew?* Morton Grove, Ill.: Albert Whitman & Co., 1995.

This book uses a pygmy shrew to make comparisons of very small objects, focusing on the molecular level. Ratios and proportions with small objects are possible avenues of study, including the magnification ratios of microscopes.

15

Proportional Reasoning: One Problem, Many Solutions!

Suzanne Levin Weinberg

PROPORTIONAL reasoning is difficult for most children, especially for those who do not understand what is actually meant by a particular proportional situation or why a given solution strategy works (Cramer and Post 1993; Lesh, Post, and Behr 1988). To begin, it is important to define exactly what constitutes a proportion. In textbooks, the word *proportion* is defined as a statement of equal fractions or equal ratios and is written as follows: $a/b = c/d$ (Levin 1999). In a popular mathematics dictionary (James and James 1992), proportion is similarly defined.

Consider the proportion $5/6 = x/24$. Equations of this type can be found in various places in the mathematics curriculum. In reality, students have seen proportions such as this as early as grade 3 under the guise of equivalent fractions. When students encounter prealgebra, most are taught to solve this equation by isolating the variable x—this can be done by multiplying both sides of the equation by the same number. Textbooks describe a completely different strategy when the term *proportion* is introduced: proportions are to be solved by cross multiplication. Figure 15.1 summarizes three ways that students might solve the equation $5/6 = x/24$.

Equivalent Fractions	One-Step Equation	Cross Multiplication
$\dfrac{5}{6} = \dfrac{x}{24}$	$\dfrac{5}{6} = \dfrac{x}{24}$	$\dfrac{5}{6} = \dfrac{x}{24}$
Find out how many times 6 "goes into" 24.	To get x by itself, undo $x \div 24$ by multiplying both sides by 24.	Cross multiply: $5 \cdot 24 = 6 \cdot x$.
6 goes into 24, 4 times.	$\dfrac{5}{6} \cdot 24 = \dfrac{x}{24} \cdot 24,$	Divide both sides by 6.
Multiply 4 by the numerator 5 and get $x = 20$.	and we get $x = 20$.	$120 \div 6 = 6x \div 6$, so $x = 20$.

Fig. 15.1. Three ways to solve $5/6 = x/24$

Solving a proportion using any of these three methods, however, does not necessarily help students understand when or how to apply proportional reasoning to appropriate situations or even why cross multiplying helps solve a proportion. Teachers have been urged to focus students' attention on the meaning of problems and to help students value different mathematically correct solutions to a single problem (NCTM 1989, 1991, 2000).

ONE PROPORTIONAL SITUATION, MANY SOLUTION STRATEGIES

On a certain map, the scale indicates that 5 centimeters represents the actual distance of 9 miles. Suppose the distance between two cities on this map measures 2 centimeters. Explain how you would find the actual distance between these two cities.

As part of a larger study on fractions and division, this question was asked of 387 middle school students (128 sixth graders, 144 seventh graders, 115 eighth graders), all of whom had been exposed to solving proportions by cross multiplication. Just 90 (≈23%) students correctly answered this problem (14 sixth graders, 39 seventh graders, and 37 eighth graders), using a variety of solution strategies (Levin 1999). A description of each of these solution strategies follows.

Find a Unit Rate

In this solution, the unit rate represents the number of miles equivalent to one centimeter; doubling allows one to find the number of miles equivalent to two centimeters. Forty-three students showed work illustrating the unit-rate strategy, and fourteen of these forty-three included an explanation for why the work was appropriate. Two of these examples appear below:

- "I divided nine by five & came out w/ 1.8. So every centimeter is 1.8 miles. So 2 centimeters equals 3.6 miles." (grade 7)

- "Divide $5\overline{)9}^{\,1.8}$ that's only 1/2 the way so 1.8 + 1.8
  $\phantom{5\overline{)9}}\,\underline{5}$ equal your anser [*sic*] so 1.8 + 1.8 = 3.6
  $\phantom{5\overline{)9}}\,40$ $\boxed{3.6 \text{ miles}}$." (grade 6)

Repeated-Subtraction Strategy

Three students used a variation of the find-a-unit-rate strategy, whereby once the unit rate was found, it was successively subtracted from the whole. For this particular problem, the number of miles corresponding to the unit rate for a single centimeter (1.8 miles per centimeter) is repeatedly subtracted from the initial number of miles.

"9 – 1.8 = 7.3;   7.3 – 1.8 = 5.5;   5.5 – 1.8 = 3.7."
(grade 7)

The student's reasoning is mathematically valid, even though a computational error was made. Without the error, the student would have arrived at the correct answer.

1.8 miles   corresponds to 1 centimeter and
9 miles   corresponds to 5 centimeters.
So,
9 – 1.8 – 1.8 – 1.8 = 3.6 miles   corresponds to 5 – 1 – 1 – 1 = 2 centimeters.

Equivalent Fractions Strategy

Eight students solved the problem by using a string of operations that correspond to the steps used to rewrite equivalent fractions. One wrote this:

"I would divide 5 by 2, giving me 2.5, then divide 9 by 2.5. That would give me the new scale 2 cm = 3.6 m." (grade 7)

This student does not explain why the sequence of steps works. The reason this sequence works can be seen by thinking about how we teach students to make equivalent fractions. Begin with the proportion involving equivalent rates: 9 miles/5 centimeters = x miles/2 centimeters. To solve this, we need to divide both 9 miles and 5 centimeters by a form of 1 that yields a fraction whose denominator is 2. We teach students to think: "5 divided by what gives 2?" The answer is 5 ÷ 2 = 2.5. Then, 9 divided by 2.5 gives x; so 9 ÷ 2.5 = 3.6 miles. This is precisely the reasoning we use to teach students to rewrite equivalent fractions when the first fraction has a larger denominator than the second—the only difference is that with equivalent fractions, we rarely multiply or divide the numerator and denominator by fractions.

Size-Change Strategy

Just eight students wrote a sequence of steps that reflects the thinking that multiplying by a ratio "stretches" or "shrinks" a given amount proportionally. Since the original number of centimeters is 5 and the desired number of centimeters is 2, the change of size involves a "shrink"; hence, the ratio

2 centimeters/5 centimeters, or 2/5, is used. Three students provided an explanation for a size-change strategy like this one:

> "Well… 5/5 of 5 cm = 9/9 of 9 mi, so 2/5 of 5 cm = 2/5 of 9 mi, so 2 cm = 3 3/5 mi." (grade 6)

It should be noted that this reasoning allows students to solve for a variable in a one-step equation of the form $x/9 = 2/5$. When a student multiplies both sides of the equation by 9, the result is $x = (2/5) \cdot 9$, or 3.6.

Cross Multiplication Using Equal Rates or Ratios Strategy

Twenty-seven students answered this question by setting up and solving a proportion using cross multiplication. Of these, six students furnished a rationale for why using a proportion is appropriate. Two students' responses are shown below. The first answer illustrates an appropriate rationale; the second illustrates how the student set up the proportion. Both students used the proportion $5/9 = 2/n$ to solve this problem.

- "5 is to 9, as 2 is to ? $5/9 = 2/n$ 3.6 miles." (grade 7)
- "I would use cross-products. Put 5 cm on top of the first fraction and put 9 miles on the bottom. Then, put 2 cm on top of the second fraction and multiply the nine miles times the two cm and divide 5 cm into the product." (grade 8)

Altogether, there are four different proportions that can be used to represent this situation. As shown below, setting up equal rates corresponds to the find-a-unit-rate strategy and setting up equal ratios leads quite naturally to the size-change strategy and equivalent-fractions strategy. Hence, setting up proportions to model a proportional reasoning situation can be used to (a) reinforce the understanding of the problem as well as (b) a means for solving the problem.

Equal rates:

$$\frac{cm}{miles} = \frac{cm}{miles} \rightarrow \frac{5 \text{ cm}}{9 \text{ miles}} = \frac{2 \text{ cm}}{x \text{ miles}} ; \quad \frac{miles}{cm} = \frac{miles}{cm} \rightarrow \frac{9 \text{ miles}}{5 \text{ cm}} = \frac{x \text{ miles}}{2 \text{cm}}$$

Equal ratios:

$$\frac{cm}{cm} = \frac{miles}{miles} \rightarrow \frac{5 \text{ cm}}{2 \text{ cm}} = \frac{9 \text{ miles}}{x \text{ miles}} ; \quad \frac{cm}{cm} = \frac{miles}{miles} \rightarrow \frac{2 \text{ cm}}{5 \text{ cm}} = \frac{x \text{ miles}}{9 \text{ miles}}$$

THINKING PROPORTIONALLY?

For the most part, students did not include any reasonable rationale for the work (either correct or incorrect) that they showed for this particular proportional situation problem. One reason might be that students are encouraged to show work but are not encouraged to explain why the steps lead to a correct answer. In this study, of the ninety students who arrived at the correct answer, only thirty-nine provided an explanation for why their strategy was correct. The remainder either showed their work but gave no additional explanation or merely described the mathematical steps they performed. The majority of the students who supplied an incorrect answer wrote no explanation of why their strategy should be appropriate, nor did their answers reflect a recognition that this problem called for proportional reasoning.

Error Pattern #1: Using a Single-Step Solution

Seventy-seven students wrote a solution involving a single operation and two of the three numbers given in the problem. A sample of these types of answers appears below:

- "You take the two cenimeters [*sic*] divide it by five." (grade 7)
- "By adding on 9 miles to each 5 centimeters." (grade 7)
- "I would subtract 5 from 9" (grade 8)
- "I would divid [*sic*] 2 by 9 and get 4.5 or 4 and a half miles." (grade 7)
- "times 2 by 9 = 18 m." (grade 7)

These answers indicate a limited understanding of the problem. As stated, each of the three numbers in the problem plays a role. Two of the numbers (5 centimeters and 9 miles) describe an initial situation. The third (2 centimeters) describes the outcome situation of one of the quantities. A one-step solution, such as the ones shown above, indicates that these students do not recognize the role of at least one of the numbers in relation to the stated problem. A failure to understand the need for all three numbers in this situation will necessarily prevent students from using a proportional reasoning approach. Furthermore, some of the answers indicate the understanding that division and multiplication are involved in some way—this could indicate partial understanding of the problem—for example, the notion of a rate. However, other answers involve operations that are in no way relevant to this problem. It is likely that at least some of these answers reflect the thinking of students who had no understanding of the problem but who wrote their best guess.

Error Pattern 2: Three Numbers Used, but in the Wrong Order

A second reason for arriving at an incorrect solution might be that students do not know why a particular method is *appropriate* for a given situation. This is a different type of misconception from that found in the one-step solutions. Examples of students' responses involving this error pattern appear below:

- "Divide the five centimeters by 2, and multiply your result by 9. 22 1/2 miles." (grade 8)
- "First I would devide [*sic*] 5 by 9 and get an answer then whatever the answer is I would times it by two." (grade 8)
- "9 – 5 = 4   4 + 2 = 6 miles." (grade 6)
- "Find 5 ÷ 9 = 0.= 0.6. From this I know that in 1 cm there is about 0.6 miles. Multiply it by 2 (from the two centimeters) and you get 1.2 the distance between the two citys [*sic*] is 1.2 miles." (grade 8)

These answers illustrate that some students recognize that all three numbers must be used, but they may not understand the role each number plays in the solution. The error shows up when the numbers are not applied "in the right order," meaning perhaps that a student knows a solution but not why it works. Again, few of the answers that are categorized Error Pattern 2 contained explanations about why the sequence of steps was appropriate.

Still a third reason for errors might be that students are "on the right track"—that is, they somewhat understand the problem—but do not know how to use mathematics to solve it. Several students wrote answers whose solutions, if they had substituted correctly, would have produced the correct answer. Other students wrote answers that are technically incorrect but show that the student does understand the proportional nature of the problem. The answers below reflect this partial understanding:

Grade 6

m.	1	2	3	4	5	6	7	8	9
c.	1	$1\frac{1}{2}$	2	$2\frac{1}{2}$	3	$3\frac{1}{2}$	4	$4\frac{1}{2}$	5

- "Find out how many miles are in each centimeters. When you find out multiply it by two." (grade 7)
- "If you look at 2 it is less then half of 5 centimeters. It is off by .5. So if you take half of 9 which is 4 1/2   2 is of [*sic*] by a half of half of 5. So take the half off 4 1/2 and your answer is 4 miles." (grade 6)

IMPLICATIONS

As mathematics teachers, we know that it is imperative for students not only to find the answer but also to determine what they did mathematically and why (see Barnett, Goldenstein, and Jackson [1994]; Cramer and Post [1993]). Although some textbooks emphasize the cross-multiply strategy for solving proportions, they rarely explain why it should be used (Levin 1999). We must help students recognize that many proportional reasoning strategies are appropriate and valid, such as the find-a-unit-rate, size-change, and equivalent-fractions strategies. We must not only teach these strategies to our students but also reinforce the underlying meaning of the proportional reasoning in the situation. If we help students learn multiple approaches to solving proportional reasoning problems, we accomplish three things: (1) We focus attention on connecting mathematics with the real world; (2) We focus attention on connecting different concepts and skills within mathematics; (3) We reinforce the themes of problem solving, communication, and reasoning by identifying the strengths and weaknesses of different strategies.

REFERENCES

Barnett, Carne, Donna Goldenstein, and Babette Jackson. *Fractions, Decimals, Ratios, and Percents: Hard to Teach and Hard to Learn.* Portsmouth, N.H.: Heinemann, 1994.

Cramer, Kathleen, and Thomas Post. "Making Connections: A Case for Proportionality." *Arithmetic Teacher* 40 (February 1993): 342–46.

James, Glenn, and Robert James. *Mathematics Dictionary.* 5th ed. New York: Van Nostrand Reinhold Co., 1992.

Lesh, Richard, Thomas Post, and Merlyn Behr. "Proportional Reasoning." In *Number Concepts and Operations in the Middle Grades,* edited by James Hiebert and Merlyn Behr, pp. 93–118. Reston, Va.: National Council of Teachers of Mathematics, 1988.

Levin, Suzanne W. "Fractions and Division: Research Conceptualizations, Textbook Presentations, and Student Performances." Doctoral diss., University of Chicago, 1998. Abstract in *Dissertation Abstracts International* 59 (1999): 1089A.

National Council of Teachers of Mathematics (NCTM). *Curriculum and Evaluation Standards for School Mathematics.* Reston, Va.: NCTM, 1989.

———. *Professional Standards for Teaching School Mathematics.* Reston, Va.: NCTM, 1991.

———. *Principles and Standards for School Mathematics.* Reston, Va.: NCTM, 2000.

16

Using Representational Contexts to Support Multiplicative Reasoning

Laura B. Kent

Joyce Arnosky

Judy McMonagle

THE development of proportional reasoning is one of the most interesting and challenging aspects of children's mathematical thinking. Proportional and multiplicative reasoning is basic to many important mathematics concepts, from simple ratios and rates to complex topics involving slope. We shall focus on specific contexts that we have found to be useful in supporting the growth of students' proportional reasoning. We use examples from our work with elementary and middle grades students as well as staff development workshops that we have conducted with classroom teachers. These examples illustrate the importance of not only contexts but also representations of those contexts in promoting students' thinking about ratios and proportions.

One of the most difficult transitions for elementary and middle level students is from additive to multiplicative reasoning. Doubling, tripling, and counting by tens are some of the strategies used by elementary grades students to multiply numbers in problem-solving situations. These same strategies can often times be used to solve proportion problems.

The problem in figure 16.1 is an example of a proportion problem. One possible strategy that children might use to figure out how many bottles are in one case of seltzer would be to divide 125 by 5 to get the answer of 25. Another possible strategy is to fill in the unfilled columns of the table using the information provided in the known columns. For example, as the number of cases increases or decreases by 1 the number of bottles increases or decreases by 25. The subsequent columns can be filled in until the number

Camp's Seltzer comes in a larger case that holds more than 15 bottles. Trudys' table was ruined when she spilled water on it. The table below shows the few numbers that Trudy can still read. Can you tell from the table how many bottles there are in each case? Explain your answer.

New Cases				5	6					12
Bottles				125	150					300

Fig. 16.1. Problem from "Number Tools" (Encyclopaedia Britannica Educational Corp., National Center for Educational Research, and the Freudenthal Institute 1997a). Reprinted with permission from the Mathematics in Context program. © 2001 Encyclopaedia Britannica, Inc.

of cases is equal to 1, which in turn tells how many bottles are in each column.

Initially, students concentrate on the additive nature of the table, that is, subtracting 25 bottles for each case that is subtracted. However, when presenting this problem to both students and teachers, we focus not only on the actual strategy but also on why the strategy with the table works. This is where the transition to multiplicative reasoning occurs. For example, we may ask, "Why is the fourth column 100 bottles instead of 124 bottles?" In other words, "Why didn't we subtract 1 from the 125 since we subtracted 1 from the 5?" Both students and teachers alike give us strange looks about why we are asking such a question but immediately respond, "That's easy, because we are talking about bottles per case!"

REPRESENTATIONAL CONTEXTS

In this example, the "bottles per case" is a context with which many children are familiar. It is also a representation of a multiplicative relationship or a ratio of bottles to cases. Thus a representational context has been developed. Ball (1993) defines a *representational context* as "the ways in which teacher and learners use the particular representation, how it serves as a tool for understanding in their work" (p. 161). What makes this example powerful is the visual image that the context of bottles per case evokes. It allows for additive reasoning to find the solution but it also "encompasses the terrain for investigation.... as well as the meanings and discourse it makes possible." (Ball 1993, pp. 160–61).

The following examples of students' work illustrate ways in which we have seen children use representational contexts to support their solutions to proportion problems. The problems are from the number strand of the NSF-funded middle grades curriculum *Mathematics in Context: A Connected Curriculum for Grades 5–8*. We shall conclude with a discussion of how we use the same problems and samples of students' work to help teachers better understand how students' proportional reasoning develops.

THE RECIPE CONTEXT

Recipe problems are useful contexts for eliciting proportional-reasoning strategies. Increasing or decreasing the number of servings of recipes and organizing the results of these computations in tables similar to the one used in figure 16.1 are processes that can also be used to help students advance their proportional-reasoning strategies. However, children do not necessarily need to be given the table to work with in order for the recipe problem to provide a representational context for proportional reasoning. For the following problem, fifth-grade students had the opportunity to construct their own representations.

> To make Sloppy Joe sandwiches for 3 people, Bob needs a half pound of ground beef. If Bob decides to make enough Sloppy Joe sandwiches for 15 people, how many pounds of ground beef will he need?

Without the structure of a table, students still used a representation of the context to construct solutions to this problem. Figure 16.2 shows examples of two additive strategies for the problem. Unlike typical rate problems such as miles per hour, the context of people and pounds of ground beef involves the association of two discrete sets. This problem also

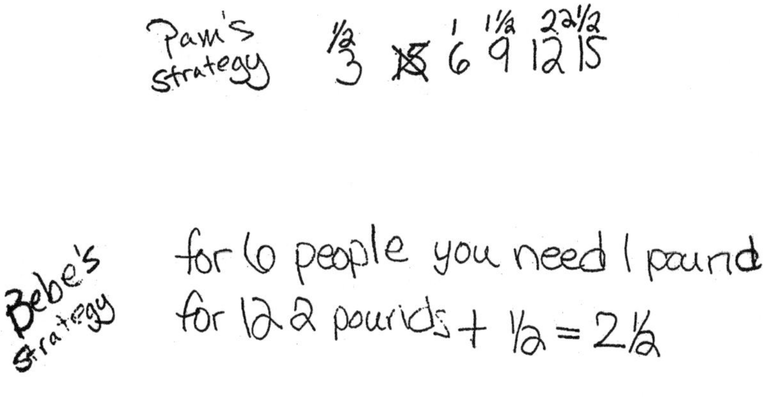

Fig. 16.2. Additive strategies for solving "sloppy joe" problem

encourages students to use a building-up strategy to solve the problem. Pam used a repeated-addition strategy to determine the amount of ground beef needed for 15 people. For each 3 people added to the number of servings, 1/2 pound of ground beef is added to the recipe. Her strategy also uses a representational structure that is very similar to the ratio-table approach in figure 16.1. Bebe used a doubling-and-adding strategy to find the answer. In both cases, as well as in the bottled-water example, the fixed relationship of the ratio is maintained as part of the building-up strategy.

In a classroom discussion about ratios and recipes, some of the same students made comments to illustrate their use of the context to reason about proportions. In the following excerpt, the students are describing the process of doubling the number of cups of chocolate chips as the number of cookies is doubled in a recipe for chocolate chip cookies.

> *Bebe:* They're the same. They are written differently, but this is still one times 36 to get 36 [cookies] and 2 times 36 to get 72. So, like Wayne was saying, each of them, the relationship is the same even though the numbers are different. The fractions look different, and what you've done is different, you've really got the same thing and that's what a ratio is…
>
> *Wayne:* If you multiply the difference in each column, it will be the same difference.
>
> *Teacher:* Say that again.
>
> *Wayne:* The difference between 1 and 36 is 35, and the difference between 2 and 72 is, um, is 70, and then 35 multiplied by 2 equals 70.

Bebe uses the context of doubling the ingredient amount as the number of servings is doubled to reflect and describe the concept of ratio. Wayne then describes a new relationship devoid of the context of recipes and describes it using proportions.

These types of responses support the argument that "situationally grounded" problems and "systematic forms of record keeping" may support the discovery of new number relations (Resnick and Singer 1993). The situation or the context of doubling the number of servings of the recipe provided the anchor for the relational thinking. The listing of the numerical relationships in the form of a ratio table provided the opportunity for students to identify new relationships by examining the existing relationships.

THE "PARKING LOT" CONTEXT

The Parking Lot problem is another example of a context that elicits representations to encourage proportional reasoning. This problem comes from the Mathematics in Context unit *Per Sense* (Encyclopaedia Britannica Educational Corp., National Center for Educational Research, and the Freudenthal Institute 1997b). As with the water bottles and recipe examples, the parking lot context, shown in figure 16.3, provides opportunities for students to make relative comparisons. The goals of these parking-lot tasks are to emphasize relative versus absolute comparisons, write equivalent fractions, and connect visual part-whole representations to mathematical symbols. In presenting this problem to her students, one teacher saw her student redraw parking P3 as shown at the top of figure 16.4 (the dots represent cars). The student shared this strategy with the class and the teacher used this drawing to help other students in the class understand the relationship between the number of spaces occupied and the total number of spaces. In addition, the drawing helped students understand the fixed relationship between the number of spaces occupied and the number of total spaces for each of the three fictitious parking lots being compared to P3 in the problem 10.

The focus on equivalent fractions and the ratio relationship helps students make the transition from the discrete context of cars and parking spaces to the "continuous bar." The continuous bar shown at the bottom of figure 16.4 is a representation that can be used specifically to emphasize the fixed relationship of a ratio since the shading of the bar emphasizes the part-whole relationship and connects fractions and percents. One strategy that we have seen students use to shade this continuous representation is first to divide the bar into 10 equal parts with each part representing 4 cars. They then shade six of the parts to represent 24 cars out of 40.

As in the previous examples, the ratio of cars to total spaces provides a useful context for the construction of meaningful representations that students can use to transition from additive to multiplicative reasoning. This context can also easily be modified to help students understand the difference between part-whole ratios and part-part ratios by changing the comparison from spaces occupied per total spaces to occupied spaces versus spaces not occupied by cars.

Once students have grasped the nature of fixed relationships through the use of discrete contexts, continuous situations such as miles per hour and other rates can then be introduced to build students' experiences with proportional reasoning. However, as Resnick and Singer (1993) cautioned, multiplicative reasoning develops slowly. What often times makes rates difficult for students to understand is the fact that rates are difficult for students to

Here are two more parking lots: P3 and P4.

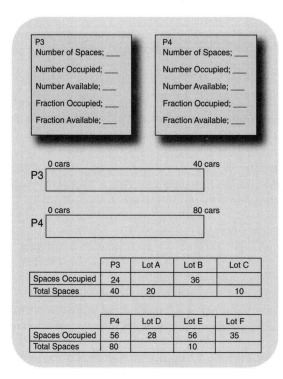

7. At first glance, which lot do you think is more occupied? Explain.

Use **Student Activity Sheet 4** to solve the following problems

8. Complete the parking lo signs for P3 and P4.

You can also make bars to show the numbers of cars in the lots.

9. Shade each bar to show the number of cars in each lot.

10. Fill in the blanks in the first table so that parkin lots A, B, and C have the same fraction of occupie spaces as P3.

Then fill in the second table so that parking lot: D, E, and F have the sam fraction of occupied spaces as P4.

This kind of table is calle a ***ratio table.*** Why do yc think it is called this?

Fig. 16.3. Parking lot problem from *Per Sense*, p. 11. Reprinted with permission from the Mathematics in Context program

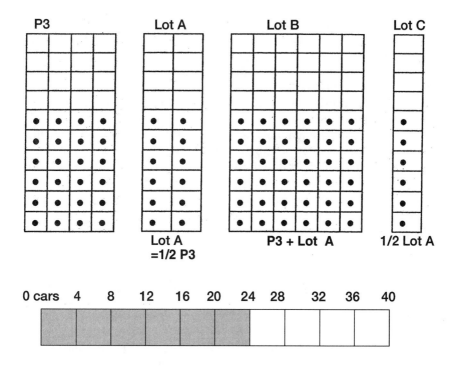

Fig. 16.4. Discrete and continuous representations of the parking-lot problem

represent. Thus, situations such as the water bottles and parking lot example should be used to help students make sense of the ratio relationship. The part-whole aspect of the continuous bar model can then be used to help them bridge the gap between counting objects to make the comparison and visualizing the relative comparison regardless of the actual numbers of objects.

CONCLUSION

Representational contexts can be used to support the development of students' multiplicative reasoning. One of the important factors to consider in selecting contexts to use is how well they lend themselves to meaningful representations. For example, the parking-lot context worked well because the cars could be reorganized in a new picture to enhance connections to continuous models that support multiplicative reasoning.

As teachers plan and sequence lessons, it is important for them to consider contexts that will help students as they transition from additive to multiplicative reasoning. The use of some rate problems such as miles per hour, although familiar to students in the sense of real-world connections, may be too abstract for students to represent. Thus, if presented with certain average speed problems prior to experiences with discrete objects, students may try to employ inappropriate strategies because the context is difficult for them to visualize and represent in a way that helps them solve the problem.

Providing organizational tools such as ratio tables can enhance students' abilities to solve a variety of rate problems. They are particularly useful if students have already experienced using them as part of their strategies with discrete situations. As concepts such as similarity and slope are introduced, the potential exists for students to continue to modify their informal strategies in favor of strategies that involve direct multiplication or division to solve proportion problems.

REFERENCES

Ball, Deborah L. "Halves, Pieces, and Twoths: Constructing and Using Representational Contexts in Teaching Fractions." In *Rational Numbers: An Integration of Research,* edited by Thomas P. Carpenter, Elizabeth Fennema, and Thomas A. Romberg, pp. 157–96. Hillsdale, N.J.: Lawrence Erlbaum Associates, 1993.

Encyclopedia Britannica Educational Corp., National Center for Educational Research, and the Freudenthal Institute. *Number Tools, Volume 1.* Mathematics in Context series. Chicago: Encyclopaedia Britannica Educational Corporation, 1997a.

———. *Per Sense.* Mathematics in Context series. Chicago: Encyclopaedia Britannica Educational Corporation, 1997b.

Resnick, Lauren B., and Janice A. Singer. "Protoquantitative Origins of Ratio Reasoning." In *Rational Numbers: An Integration of Research,* edited by Thomas P. Carpenter, Elizabeth Fennema, and Thomas A. Romberg, pp. 107–30. Hillsdale, N.J.: Lawrence Erlbaum Associates, 1993.

17

Interpretations of Fraction Division

Rose Sinicrope

Harold W. Mick

John R. Kolb

IF OUR students are to construct a rich, relational understanding of fraction division, we as teachers need a framework for fraction-division situations that will help us select problem types and to design tasks. What is fraction division? What is entailed? To answer these questions from a teaching and learning perspective, we need to know what kinds of situations are fraction-division situations, what reasoning occurs within these situations, and what mathematical generalizations can be made.

Fraction-division algorithms can arise as abstractions of procedures used to reason out the solutions to different problem situations. By exploring different algorithms, problem situations, and instructional models, we can categorize fraction-division situations, here called "interpretations."

For whole-number division, problem situations can be categorized as *measurement* division (determining the number of groups); *partitive* division (determining the size of each group); or the inverse of a *Cartesian product* (determining a dimension of a rectangular array). Fraction division can be explained by extensions of all three of these whole-number interpretations. But these extensions are not sufficient; division as the determination of a unit rate and division as the inverse of multiplication are also important fraction-division interpretations.

MEASUREMENT DIVISION

When Warrington (1997) asked her fifth- and sixth-grade students to describe what 4 ÷ 2 meant to them, the most common response was, "It means how many times does 2 fit into 4 or how many groups of two fit into four?" This measurement interpretation is a common instructional focus for whole-number division. As Warrington's students demonstrated, it is also a meaningful interpretation for fraction division. Her students reasoned that 2 divided by 1/2 is 4 because "one-half goes into 2 four times" and " if you had two candy bars and you divided them into halves, you'd have four pieces."

Instructional models like pattern blocks also use this measurement interpretation. Pattern blocks—yellow hexagons, red trapezoids, blue parallelograms, and green triangles—are used in fraction-division instruction. The pattern block pieces are constructed in such a way that two green triangles will cover a blue parallelogram, for example, and six green triangles or two red trapezoids will cover a yellow hexagon. For the convenience of fraction-division instruction, the unit is often defined as the region formed by two adjacent hexagons. This unit makes it easy to show halves, thirds, fourths, sixths, and twelfths. The problems are presented using shapes. Students may be asked, for example, to determine how many red trapezoids, 1/4 of the whole, it will take to cover 11 green triangles, 11/12 of the whole, that is 11/12 ÷ 1/4. Since 1 red trapezoid will cover 3 green triangles (1/4 = 3/12), it will take 11 ÷ 3, or 3 2/3, trapezoids to cover the 11 triangles (See fig. 17.1.).

The algorithm that represents the procedural reasoning in this type of division is the common-denominator algorithm for the division of fractions:

$$\frac{a}{b} \div \frac{c}{d} = \frac{ad}{bd} \div \frac{bc}{bd} = \frac{ad}{bc}$$

The first step in the common-denominator algorithm is to express both the divisor and the dividend as fractions with like denominators. Once the denominators (the units of measure) are the same, the numerators are divided as in figure 17.1.

It is possible to relate the procedural reasoning used in the solution of measurement divisions to the invert-and-multiply algorithm. The instructional process follows the pattern of first dividing a whole number by a unit fraction, for example, 4 ÷ 1/3. Here the reasoning is that there are 3 thirds in 1, and hence four times as many in 4, or 12 one-thirds in 4. Next, the question asked is, "How many two-thirds are in 4, or 4 ÷ 2/3?" The reasoning is that since there are 12 one-thirds in 4, then there are half as many two-thirds in 4, or 1/2 of 12 = 6; that is, dividing by 2/3 is equivalent to multiplying by 3 and then multiplying by 1/2, or 3 × 1/2 = 3/2, or multiplying by the reciprocal of 2/3.

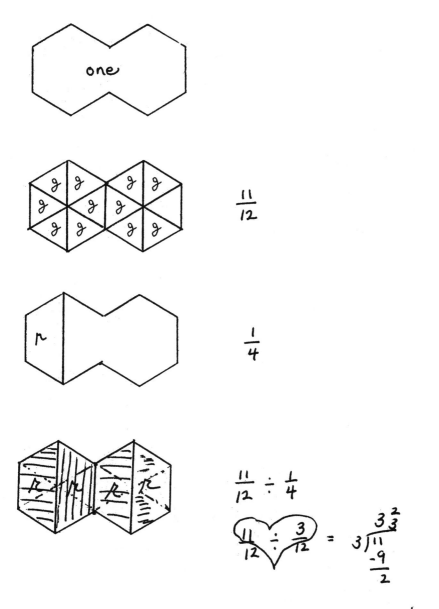

$$3\tfrac{2}{3} \text{ red trapezoids will cover 11 green triangles}$$

Fig. 17.1. 11/12 ÷ 1/4

Partitive Division

Another meaning of whole-number division that can be extended to fraction division is partitive division. The following is an example of partitive division with whole numbers that is a sharing situation.

Jo has 12 pies. She will share them equally among 3 friends. How many pies will each friend receive?

If we use the same division meaning and extend it to fractions, then the problem takes the following form:

Jo has 12 sixteenths of a pie. She will share the pie equally among 3 friends. How much of the pie will each friend receive?

Solution:
12 wholes ÷ 3 = 4 wholes
12 sixteenths ÷ 3 = 4 sixteenths

Although this is a contrived "real world" problem, understanding the situation enables us to translate real-world problems of a similar type into a fraction-division expression with an algorithmic solution that is meaningful. Perhaps in our haste to get to the invert-and-multiply algorithm, we often skip the division of a fraction by a whole number. These divisions can take two forms:

$$\frac{a}{b} \div c = \frac{a \div c}{b} \text{ when } c \text{ divides } a.$$

$$\frac{a}{b} \div c = \frac{a}{b \times c} \text{ when } c \text{ does not divide } a.$$

In her study of prospective teachers' knowledge of fraction division, Tirosh (2000) used the following partitive division in which the divisor does not divide the numerator (p. 9):

Four friends bought 1/4 kilogram of chocolate and shared it equally. How much chocolate did each person get?

The solution requires dividing 1/4 by 4. A strip of paper folded into four equal parts with one part shaded can represent the chocolate. With three sections folded back in accordion-fold style to show only the 1/4 kg of chocolate, fold the 1/4 section into four equal parts and mark the amount one person will get. Unfold the strip to name the amount of chocolate, 1/16 kg.

DETERMINATION OF A UNIT RATE

A different aspect of partitive division occurs if the situation is not about sharing but focuses on the size of one group. The following problem is an example of fraction division as the determination of a unit rate.

A printer can print 20 pages in two and one-half minutes. How many pages does it print per minute?

A possible solution:
20 pages in 2 1/2 minutes
40 pages in 5 minutes
8 pages in 1 minute

$$\frac{20}{2\frac{1}{2}} = \frac{20}{\frac{5}{2}} = \frac{20 \times 2}{\frac{5}{2} \times 2} = \frac{40}{5} = \frac{40 \div 5}{5 \div 5} = \frac{8}{1}$$

Note that the process of multiplying by 2 and dividing by 5 is equivalent to multiplication by the fraction 2/5, which provides another rationale for the invert-and-multiply algorithm.

Ott, Snook, and Gibson (1991) use the term *partitive division* for this division interpretation. For them, the determination of the unit rate is equivalent to determining the size of one set; it is the same question that is asked in the partitive division of whole numbers. Ott, Snook, and Gibson (1991, p. 9) used the following question in their explanation of the partitive division of fractions: "If 1/6 of an egg carton is to form 2/3 of a set, what is the size of one set?" They used the opposite order of dividing by the numerator and then multiplying by the denominator to determine the size of one set. Summarizing the actions, we find that dividing by 2/3 is accomplished by multiplying by 3/2 (see fig. 17.2.). With other examples, it becomes possible to generalize that dividing by a fraction is equivalent to multiplying by the reciprocal of the fraction, which is the invert-and-multiply algorithm:

$$\frac{\frac{a}{b}}{\frac{c}{d}} = \frac{\frac{a}{b} \times \frac{d}{c}}{\frac{c}{d} \times \frac{d}{c}} = \frac{\frac{ad}{bc}}{1} = \frac{ad}{bc}$$

Step 1.

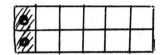

$\frac{1}{6} \div \frac{2}{3}$

$\frac{2}{12} \div \frac{2}{3}$

Represent $\frac{1}{6}$.

$\frac{1}{6} = \frac{2}{12}$.

Step 2.

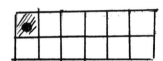

$\frac{2}{12} \div (2 \cdot \frac{1}{3})$

$(\frac{2}{12} \cdot \frac{1}{2}) \div \frac{1}{3}$

$\frac{1}{12} \div \frac{1}{3}$

If $\frac{2}{12}$ is $\frac{2}{3}$ of a set,
then $\frac{1}{2}$ of $\frac{2}{12}$ or $\frac{1}{12}$
is $\frac{1}{3}$ of the set.

Step 3.

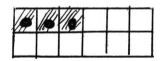

$\frac{1}{12} \cdot 3$

$\frac{3}{12}$

If $\frac{1}{12}$ is $\frac{1}{3}$ of the
set, then $3 \cdot \frac{1}{12}$ is
1 set.

Fig. 17.2. 1/6 ÷ 2/3

DIVISION AS THE INVERSE OF MULTIPLICATION

The following problem is an example of the inverse of a fraction multiplication in which one of the fractions is an operator:

In a seventh-grade survey of lunch preferences, 48 students said they prefer pizza. This is one and one-half times the number of students who prefers the salad bar! How many prefer the salad bar?

A Possible Solution:
11/2 × number for salad = 48
3/2 × number for salad = 48

If you multiply [the number for salad] by 3 and divide by 2, you get 48.

To undo, multiply by 2 and divide by 3, or multiply by the fraction 2/3.

48 ÷ 1 1/2 = 48 ÷ 3/2 = 48 × 2/3
32 prefer salad.

Although different strategies can be used to solve this problem, a general approach is reversing the procedure of the original multiplication. Multiplication by 1 1/2 is equivalent to multiplication by 3/2. To multiply by 3/2, multiply by 3 and divide by 2. The reverse is to multiply by 2 and divide by 3, or to multiply by 2/3. The symbolic representation becomes

$$\frac{a}{b} \div \frac{c}{d} = \frac{a}{b} \times \frac{d}{c}.$$

DIVISION AS THE INVERSE OF A CARTESIAN PRODUCT

Another interpretation for the division of fractions is the inverse of a Cartesian product. The problem situation may be one of area, where the total area and one dimension of a rectangular region are known. The problem is to determine the other dimension.

For example, to determine the width of a rectangle that has a length of 3/4 units and an area of 6/20 square units, first form a corner of a rectangle with two sides (see fig. 17.3.). Along one side, mark off a length of 3/4 units. The unit length is arbitrary, but put tick marks at each fourth. In order to deal with the area, which is measured in twentieths, mark off the other side in fifths, so that the area will be represented in little rectangles representing

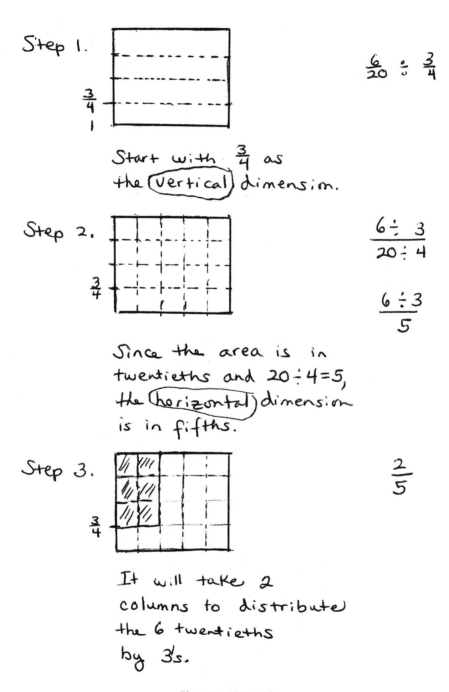

Step 1.

$\frac{3}{4}$
1

$\frac{6}{20} \div \frac{3}{4}$

Start with $\frac{3}{4}$ as
the (vertical) dimension.

Step 2.

$\frac{3}{4}$

$\dfrac{6 \div 3}{20 \div 4}$

$\dfrac{6 \div 3}{5}$

Since the area is in
twentieths and $20 \div 4 = 5$,
the (horizontal) dimension
is in fifths.

Step 3.

$\frac{3}{4}$

$\dfrac{2}{5}$

It will take 2
columns to distribute
the 6 twentieths
by 3's.

Fig. 17.3. 6/20 ÷ 3/4

1/20 of a square unit. Next, determine the width by marking off one-fifths until you use all 6/20 square units of area. Since in our rectangle each one-fifth in width uses three 1/20 square units of area, the width is 2/5 units. An algorithm that describes this process is an algorithm in which numerator divides numerator and denominator divides denominator—for example,

$$\frac{6}{20} \div \frac{3}{4} = \frac{6 \div 3}{20 \div 4} = \frac{2}{5}.$$

Summary

Fraction division has many different interpretations. For instance, we divide to determine how many times one quantity is contained in a given quantity, to share, to determine what the unit is, to determine the original amount, and to determine a dimension for an array. For the teacher of mathematics, an exploration of different interpretations of fraction division forms a framework for designing instruction through posing problems. As students solve the teacher-posed problems and similar problems, they can eventually generate algorithms for solving even "larger sets" of problems. This connection between the problem context and the fraction-division algorithm is a link that is often missing in students' understanding and performance (Piel and Green 1994). These problem contexts for fraction division are varied in ways deeper than whether the problem is about pizza or gallons of milk. These contexts encourage different procedural reasoning to solve the problems. These situations or interpretations can be categorized as measurement division, partitive division, determination of a unit rate, the inverse of multiplication, and the inverse of a Cartesian product.

References

Ott, Jack M., Daniel L. Snook, and Diana L. Gibson. "Understanding Partitive Division of Fractions." *Arithmetic Teacher* 39 (October 1991): 7–11.

Piel, John A., and Michael Green. "De-Mystifying Division of Fractions: The Convergence of Quantitative and Referential Meaning." *Focus on Learning Problems in Mathematics* 16 (Winter 1994): 44–50.

Tirosh, Dina. "Enhancing Prospective Teachers' Knowledge of Children's Conceptions: The Case of Division of Fractions." *Journal for Research in Mathematics Education* 31 (January 2000): 5–25.

Warrington, Mary Ann. "How Children Think about Division of Fractions." *Mathematics Teaching in the Middle School* 2 (May 1997): 390–94.

18

Developing Understanding of Ratio-as-Measure as a Foundation for Slope

Joanne Lobato

Eva Thanheiser

WHAT do students think they are measuring when they find the slope of a line? Consider the response of the highest-performing student in an Algebra 1 class to the task shown in figure 18.1, during an interview of the student by the first author. Marni correctly calculated the slope as 1/2 using the "rise over run" formula. However, her response indicates that she conceives of slope only as a number, not as a measure of the dripping rate of the faucet:

Interviewer: OK, what is your slope again?

Marni: A half slope.

Interviewer: A half of what?

Marni: I guess it just shows the height, I mean, umm, or the slope.

Interviewer: So, if your slope is a half, and I asked a half of what …

Marni: It isn't a half of anything, I think. It just determines the measurements on how high it is rising.

What Does Slope Measure?

Textbooks often describe slope as a measure of the direction and steepness of a line. However, if slope is a measure of the steepness of a line, then two lines with the same slope should have the same steepness, which is not the

The preparation of this chapter was supported by grant no. REC-9733942 from the National Science Foundation. The views expressed in this chapter do not necessarily reflect official positions of the National Science Foundation.

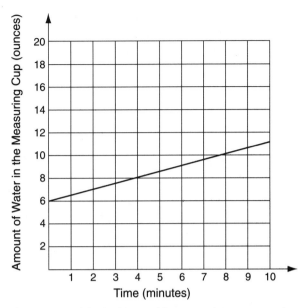

The leaky faucet task. A leaky faucet drips water into a measuring cup.
The graph shows the water level in the measuring cup over time. The cup
started out with some water in it before the dripping began. What is the
slope of the line and what does it tell you about the situation?

Fig. 18.1. The leaky faucet task

case for the example shown in figure 18.2. This difficulty can be addressed
by stipulating that slope is a measure of the steepness of lines within the
same scale system. Yet, this begs the question of whether the primary purpose of calculating slope is to obtain a measure of the steepness of a line.

Consider an alternative conception of slope as the rate of change in one
quantity relative to the change of another quantity, where the two quantities
covary. By quantity, we mean an aspect of an object or situation that can be
measured, such as length, age, or time. Slope is a measure of different attributes (like velocity, sweetness, density, or gas efficiency) depending on the
quantities involved in the relationship. For example, imagine the graph of
a line where the x-axis represents elapsed time and the y-axis represents
distance. The slope of the line represents velocity and is a measure of the
object's motion, rather than a measure of the steepness of the line.

Slope can provide a measure of the steepness of an object, where the object
is a physical object like a slide or hill (as opposed to a line). In such a situation, slope is the rate of change of the vertical distance relative to the horizontal distance. In contrast, treating slope as a measure of the steepness of a

line focuses attention on the line as a physical object rather than as an infinite collection of ordered pairs of values of covarying quantities, which can lead to the problem illustrated in figure 18.2.

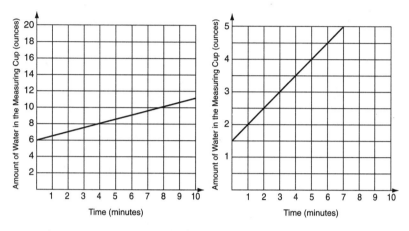

Fig. 18.2. Two lines with the same slope but not the same steepness

Slope involves the construction of a ratio as the measure of a given attribute (what Simon and Blume [1994] call a "ratio-as-measure"). Constructing ratio as a measure appears to be a complex process for at least three reasons. First, ratio-as-measure tasks are likely to require students to reason proportionally. In contrast, when working with the slope formula, students can count squares on a coordinate grid system to find a rise and run and then simply write one number on top of the other, without mentally forming a ratio. Second, ratio-as-measure tasks involve aspects of modeling, like focusing on one quantitative relationship in a complex situation involving several relationships. Third, conceiving of an indirect measure, like a ratio, is difficult when students' experiences have been dominated by the use of direct measures. For example, when asked to draw a picture of a ramp with a slope of 2, another student interviewed by the first author responded, "2 what? 2 inches? 2 millimeters?" The response suggests that the student conceives of measures as direct (e.g., where one reads a measure from a ruler). Students need help through instruction to develop an understanding of the legitimacy of indirect measures.

Purpose and Overview

We provide a general framework for helping students learn to construct ratios as measures—a foundation for understanding slope. This framework addresses the modeling and proportional reasoning aspects of constructing ratios as measures. Each component of the framework is illustrated with two examples, one from the situation of creating a ratio as a measure of the steepness of a wheelchair ramp and the other from the situation of creating a ratio as a measure of how fast an object travels. By providing examples from multiple contexts, we hope to illustrate a general approach to the process of creating ratios as measures.

Instructional Settings

The examples are drawn from two teaching experiments, both taught by the authors. The first teaching experiment involved nine high school students for thirty hours of instruction during a summer. The second teaching experiment is ongoing and involves six high school students who meet after school once a week. In both cases, we selected average-performing students (not the very top students and not those who were completely lost)—usually students earning B's or C's. The instruction involved two computer environments, Geometer's Sketchpad (GSP) (1996) and MathWorlds (developed for the SimCalc project, directed by James Kaput) (1996). The conditions in both teaching experiments were intentionally designed to be more ideal than what teachers typically encounter, since the experiments were part of a larger research study designed to investigate learning under various conditions. However, we believe that the approach represents a promising first step, which can serve as a source of stimulation for teachers who wish to modify or extend the activities to fit the needs of their classrooms.

FOUR COMPONENTS OF UNDERSTANDING RATIO AS MEASURE

1. Isolating the Attribute That Is Being Measured

Rather than beginning with slope formula instruction and then applying slope to real-world situations, we started with situations and posed "measure" questions that could eventually be addressed by constructing slope. We quickly learned that when we asked students how they would measure the steepness of a wheelchair ramp, they did not automatically measure the height and base of the ramp and form a ratio, even though the students had received slope instruction in school. Similarly, when students were asked to measure how fast a person walks, they rarely measured both distance and time.

One unexpected source of difficulty was isolating the characteristic of steepness in the wheelchair ramp situation and the characteristic of "fastness" in the walking situation. The reason for this difficulty may be that these attributes are not the only attributes involved in each situation and may not be the most salient characteristics for students who have had relevant everyday experiences with inline skating ramps, stairs, races, and walking. Thus, students need help isolating the different attributes.

Example A: Sorting out steepness from "work required to climb" in the ramp situation

The students in the summer teaching experiment had developed meaning for steepness as the "slantiness" of a hill or ramp and could draw one hill that was steeper than another. However, two difficulties emerged, which indicated that students had trouble isolating the attribute of steepness from other attributes in the situation.

First, when asked if a hill was the same steepness throughout, two-thirds of the students thought that the hill became steeper as they imagined climbing up the hill. Their reasons varied: the angle changes, the height increases, or you walk a longer distance. The following interchange illustrates one student's reasoning:

> *Teacher:* Where is it steeper?
>
> *Terry:* Like right there [points to the far right of the hill]
>
> *Teacher:* OK, and how do you know it's steeper right there?
>
> *Terry:* 'Cause it's like higher up on the angle.
>
> *Teacher:* OK. So being higher up on the angle makes it steeper?
>
> *Terry:* Uh, hmm.
>
> *Teacher:* So is the steepness constantly changing?
>
> *Terry:* Yeah.

Second, when the teacher asked the students to draw two nonidentical hills with the same steepness, a debate emerged over Jim's drawing (see fig. 18.3). Many students argued that the hill on the right was steeper because it was higher, longer, or harder to climb. The most salient feature of the situation for the students may have been that people become more tired when climbing the hill on the right. If this attribute of "work required to climb" is not isolated from the attribute of steepness, then students may incorrectly conclude that the hills in figure 18.3 aren't the same steepness or that a single hill becomes steeper near the top (since a person becomes more tired as he or she continues to climb).

Fig. 18.3. Jim's drawing of two hills with the same steepness

One way to respond to this difficulty is to ask students to make a list of the ways that the two hills in figure 18.3 are alike and different. By discussing the differences as well as the similarities, students in the teaching experiment were eventually able to isolate steepness from other attributes in the situation.

Example B: Sorting out "motion through space" from "leg motion"

A similar problem developed with the motion situation. Students watched a computer simulation in MathWorlds of two characters walking toward each other at different speeds. The students were asked to decide which character walked faster and how they might measure how fast each character was traveling. (Distance and time were not displayed.)

All three pairs of students in our ongoing teaching experiment thought that the number of steps would affect how fast a character was going. Bonita and Carissa thought that the two characters went the same speed because they both walked the same number of steps in the same time (i.e., their legs were moving at the same rate). They did not consider the size of the steps or the distance traveled. Similarly, Isaac noticed that the characters' feet were moving at the same pace and incorrectly concluded that the characters were going the same speed. We concluded that asking how fast someone travels is ambiguous since there are at least two different types of "fastness": (1) how fast an object moves through space, and (2) how fast the object's legs move. Although we intended for students to focus on the first attribute, many focused on the second. One reason why students focused on leg motion appears to stem from a common, everyday experience of a child trying to keep up with an adult. As Adolfo put it, "A child would have to go fast and the mother slow so that they would go at the same pace."

To help students disentangle the two rates—speed versus the rate of leg motion—the teacher pretended to be a robot that could be "programmed." She asked students to tell her what to do so that she would walk fast, but she did not allow the students to use the word "fast" in their instructions. Carissa suggested that the teacher walk seven steps, reasoning that seven was a pretty big number and would force the teacher to walk quickly. When the teacher took seven large steps very slowly, Carissa was surprised and told the teacher to move her legs more quickly. The teacher followed Carissa's instructions but took seven tiny steps in a short period of time, which made her feet move quickly but her body move slowly through space. This event was counter to Carissa's expectation. By resolving the conflict that she encountered, Carissa was eventually able isolate "motion through space" from "leg motion" and realize that the number of steps taken is insufficient to determine how fast a person is moving through space.

2. Determining Which Quantities Affect the Attribute and How

Once students have isolated an attribute like steepness or motion through space, then they need to determine which quantities affect that attribute. A previous study demonstrated the complexity of determining the effect of changing quantities on an attribute (Lobato and Thanheiser 1999). When asked to create a measure of the steepness of a wheelchair ramp (as shown in fig. 18.4) nearly half of the high school students were uncertain about the role of the platform (e.g., they didn't know whether increasing or decreasing the length of the platform would change the steepness of the ramp). Furthermore, most of the students had a harder time understanding the effect of changing the length of the base (minus the length of the platform) than the effect of changing the height. For example, one student argued that if you make the base shorter, then the ramp will become less steep, and if you make the base longer, then the ramp will become steeper. The teaching experiments were designed to address these difficulties.

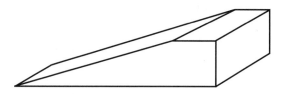

Fig. 18.4. Diagram of a wheelchair ramp

Example A: Determining which quantities affect the steepness of a wheelchair ramp

Students were asked to list all the quantities they saw in the wheelchair ramp in figure 18.4. Then the class selected quantities one by one, figured out a way to change the quantity, and observed what happened to the steepness of the ramp. For example, to test the effect of changing the width, Nathan constructed two paper ramps that varied by width only. Since the skinny and wide ramps had the same steepness, Nathan concluded that width doesn't affect steepness.

GSP was used to explore other quantities (fig. 18.5). Students moved quantities (like the length of the platform and the height of the ramp) and noted changes in the steepness of the ramp. They concluded, among other things, that decreasing the base of the ramp made the ramp steeper but increasing the base made it less steep.

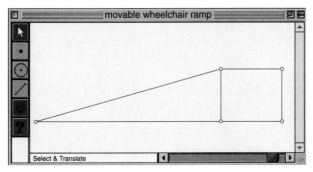

Fig. 18.5. Dynamic sketch of a wheelchair ramp in Geometer's Sketchpad (1996)

Example B: Determining the effect of time and distance on speed

In the second teaching experiment, students explored how changing time and distance affected the motion of an object by entering distance and time values for various characters in MathWorlds (using a special script written for our project by Jeremy Roschelle and Janet Bowers, as shown in fig. 18.6). In one problem, students were asked to enter a time and a distance for the frog so that it would travel slower than a clown that traveled 15 cm in 6 sec. Students experienced several difficulties. For example, Priyani and Adolfo kept returning to the idea that speed could be measured by time alone—an idea that has roots in the everyday experience of determining the speed of runners. Since the distance in a race is fixed, one can measure how fast a runner travels by simply timing the race.

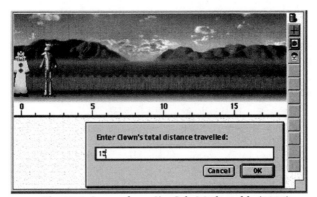

Fig. 18.6. Screen from SimCalc Mathworlds (1996)

As a result of associating speed with time, Priyani and Adolfo could easily argue that they could make the frog go slower by increasing the time and faster by decreasing the time. However, when asked to give both characters the same time and change only the distance to make the frog go slower, the students did not know what to do. They thought that having the frog walk a greater distance would make him go slower. Thus, they were surprised to run the simulation and see that this was not the case. The simulation helped them determine how changes in distance affect speed.

3. Understanding the Characteristics of a Measure

There are several characteristics of a good measure, which students are often aware of only implicitly. One characteristic is the legitimacy of indirect measures. A second characteristic we shall call "reproducibility." Imagine a ramp with a height of 8 cm and a base of 4 cm, and thus a slope of 2. One reason that slope is a good measure of the steepness of a physical object is that someone else can create a second ramp with a slope of 2 and the two ramps will share the same steepness, even though the ramps may not be identical (e.g., the second ramp may have a height of 10 cm and base of 5 cm). Thus the attribute of same steepness is reproduced in the second ramp because of the measure that is used.

Example A: Measuring steepness

After the students had explored some of the quantitative relationships in the wheelchair ramp, the teacher asked how they could measure the steepness of the ramp. Denise suggested that they use a ruler to measure "the part you walk up," which was apparently an attempt to measure directly the part of the apparatus that is steep. She found that the length of the ramp in the drawing was 6 inches. In an effort to establish the importance of the characteristic of "reproducibility," the teacher explained that if 6 inches is to be a good measure of the steepness of a ramp, then the students should be able to draw nonidentical ramps with a steepness of 6 inches and obtain ramps with the same steepness. When the students carried out this activity, they discovered that some of their ramps were steeper than others. Thus, they concluded that the length of the ramp alone was not a good measure of steepness.

This process was repeated as the students suggested and tested the following possible measures of steepness (which were all direct measures): the height of the ramp, the length of the base, and the angle of inclination. After testing their measures, the students rejected height alone and length alone as measures of steepness but were pleased to discover that the angle worked. The teacher validated their discovery, but she also told the students that people in many jobs (like construction) also use a different measure. Thus, she asked them to continue working on this activity to come up with a second measure of steepness.

Katie began measuring the lengths and heights of the two ramps with the same steepness. Because of her imprecise measurements, Katie was unable to see that the lengths and heights of the two ramps were proportional. However, the focus on height *and* length allowed the teacher to suggest that the students should consider two quantities rather than one in their measure. The teacher then posed an activity (to be described shortly) that led to the construction of ratio for many of the students.

Example B: Measuring motion

When students were asked to create a measure of how fast an object travels, several students used direct measures, like time. One way to help students see that time is insufficient is to ask two students each to run for five seconds. Seeing that it is possible for one student to run faster than the other can help students understand that time alone is not a good measure of speed. We found that helping students test their direct measures often led to an understanding of the inadequacy of the direct measure and created the need for an indirect measure.

4. Constructing a Ratio

Once students have identified two quantities to include in their measure of an attribute (e.g., height and length in the ramp situation or distance and time in the speed situation), students still need help to move from considering the measure as two whole numbers (e.g., a character walks 10 cm in 4 sec.) to understanding the measure as a ratio (e.g., 10 cm for every 4 sec., which is equivalent to 2.5 cm each sec.). Computer environments can play an important role in the construction of a ratio by helping students conceive of a situation dynamically. We found that the computer also allows students to generate a family of values with a given attribute (e.g., 10 cm in 4 sec. is the same speed as 30 cm in 12 sec.) by guessing and checking or using numeric strategies. Reflection and discussion can then help students construct a ratio as a measure.

Example A: "Same steepness" activity

Students were asked to create as many ramps as possible with the same steepness as a ramp with a height of 3 cm and length of 12 cm. Brad used GSP to create the family of ramps shown in figure 18.7.

Interestingly, Brad discovered the "slope" function in the "measure" menu of GSP and found that the slope of one of his ramps was 0.25. He told the teacher that he thought that steepness was slope, and he remembered from

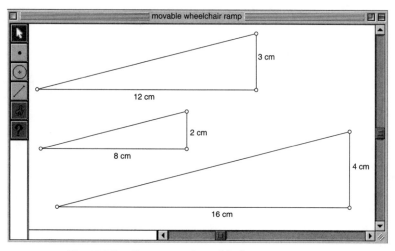

Fig. 18.7 Brad's construction with the same steepness

school that slope was "rise over run." Although he correctly identified the "rise" and "run" of his ramp, he admitted that he had no idea what the 0.25 meant in the ramp situation. The teacher suggested that Brad temporarily set the 0.25 aside, create more ramps with the same steepness, and look for patterns in the height and length measurements. After about 15 minutes, Brad noticed that the length of each ramp was four times as large as the height. When the teacher asked how the height compared to the length, Brad figured out that each height was 1/4 (or 0.25) of the corresponding length. He constructed meaning for the slope as the multiplicative comparison of height to length. It is interesting to note that this construction of ratio developed, not by memorizing a formula, but by mentally comparing the heights and lengths of a set of ramps that shared the same steepness.

To help Brad also think about the relationship of change between the height and length, the teacher asked him to do two followup activities: (1) create a ramp with the same steepness as the other ramps but with a length of 1 cm, and (2) explain how you would change the height of the original ramp if you increased its length from 12 cm to 13 cm but wanted the new ramp to have the same steepness as the original. The purpose of these questions was to help develop the idea that for every increase of 1 cm in length, there will be a corresponding increase of 1/4 cm in height.

Example B: "Same speed" activity

In the summer teaching experiment, students were asked to create a table of values that would make the frog in the Mathworlds environment (fig. 18.6) walk the same speed as a clown walking 10 cm in 4 sec. Many students used a "guess and check" strategy (e.g., one student tried 15 cm and 8 sec. and then kept adjusting the time until he arrived at 15 cm in 6 sec.). Other students identified numeric patterns by reflecting on their initial entries (e.g., some students found that multiplying the distance and the time by the same whole number resulted in values that worked). However, no one was able to explain why the numeric patterns worked.

A class discussion quickly showed the limitations of the numeric strategies. Brad shared his solution of 90 cm in 35 seconds, which he found by guessing and then running the simulation. Terry disagreed, arguing, "Ten goes into 90 nine times, and 4 goes into 35 eight times and a little bit left over." But the other students admitted that they couldn't follow Terry's (numeric) explanation, and Terry did not offer a better explanation at that time.

The teacher asked the students to explain why a simpler problem involving doubling worked (i.e., why walking 20 cm in 8 sec. was the same speed as walking 10 cm in 4 sec.). Terry and Jim used Terry's drawing (see fig. 18.8) to explain that for both characters to have the same speed they would need to walk 10 cm in 4 seconds at the same time. This explanation fits with the students' experience of watching the computer simulation to see if the characters were walking "neck-and-neck" during the first 10 cm and ignoring the rest of the frog's journey. However, neither student explained why 20 cm in 8 seconds "worked" or accounted for the relationship of the remainder of the frog's journey to his speed, which might explain why the drawing did not show the frog's distance as twice the clown's distance.

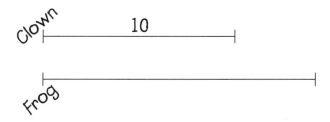

Fig. 18.8. Author's re-creation of Terry's nonproportional drawing of the distances 10 cm and 20 cm

A breakthrough occurred when Brad appeared to construct a ratio. Brad explained that doubling works "because the clown is walking the same distance; it's just that he's walking the distance twice ... he walks 10 cm in 4 seconds ... he's going to walk it again, another 4 seconds, another 10 cm."

The creation of the "10 cm in 4 seconds" unit or ratio was adopted by other students and combined with partitioning (i.e., subdividing the unit)

and iterating (i.e., repeating the unit) to create more powerful explanations. For example, Terry eventually was able to explain why walking 2.5 cm in 1 second was the same speed as walking 10 cm in 4 seconds. He stated that "it would be like he's walking one-fourth of the 10 and 4; it's like one-fourth of each thing" (meaning 1/4 of the 10 cm and 1/4 of the 4 seconds). He partitioned the "10 cm in 4 seconds" unit into four segments, formed a new "2.5 cm in 1 second" segment (as indicated by the circled section in figure 18.9), and then iterated the "2.5 cm and 1 second" unit four times to re-create the original "10 cm in 4 seconds" unit.

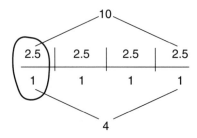

Fig. 18.9. Author's re-creation of Terry's drawing showing why 2.5 cm in 1 second is the same speed as 10 cm in 4 seconds

CONCLUDING COMMENTS

In order for slope to be meaningful to students across a broad range of situations, students need to develop an understanding of slope as a ratio that measures some attribute in a situation. However, these are many complexities associated with constructing ratios as measures. Some people may argue that slope should not be made any more difficult than the slope formula. They are right if we are satisfied when students are only able to solve textbook problems that cue students when and how to use the slope formula. In contrast, real-world situations involving rates of change are usually messier and more complex. Rather than avoiding complexity, instructional activities should help students learn how to cope with it. The examples presented here suggest ways for teachers to help students develop an understanding of the modeling and proportional reasoning aspects of ratio-as-measure tasks, which in turn can help students develop an understanding of slope that is more general and applicable.

REFERENCES

The Geometer's Sketchpad, Version 3.0. Emeryville, Calif.: Key Curriculum Press, 1996.

Lobato, Joanne and Eva Thanheiser. "Rethinking Slope from Quantitative and Phenomenological Perspectives." *Proceedings of the Twenty-first Annual Meeting of the North American Chapter of the International Group for the Psychology of Mathematics Education* 1 (1999): 291–97.

MathWorlds, Version 1.1b4. SimCalc Project, University of Massachusetts Dartmouth, 1996. www.simcalc.umassd.edu

Simon, Martin, and Glendon Blume. "Mathematical Modeling as a Component of Understanding Ratio-as-Measure: A Study of Prospective Elementary Teachers." *Journal of Mathematical Behavior* 13 (June 1994): 183–97.

19

Using Technology to Teach Concepts of Speed

Janet Bowers

Susan Nickerson

Garrett Kenehan

RECENT nationwide studies have confirmed that many students in the lower and middle grades have difficulties learning the concepts of ratio and rate (Kouba, Zawojewski, and Strutchens 1997). For example, after analyzing the results of the 1992 National Assessment of Educational Progress (NAEP) study, Kouba and her colleagues concluded that "whereas growth is evident over the grades, the overall low level of performance on percent, ratio, and rate word problems is troubling" (p. 134). What is perhaps even more significant is the message included in the authors' closing remarks (p. 138):

> The most troubling results were the low performance levels associated with students' ability to justify or explain their answers to regular and extended constructed-response items ... The 1992 NAEP results suggest that ... students will need considerably more experience with justifying and explaining their mathematical work than they have had in the past.

To address this deficit, the NAEP analysts called for research "identifying both the nature of expectations regarding communication in mathematics and the means for achieving it in mathematics classrooms" (p. 138). Although we cannot offer a "silver bullet" solution in one or two activities that will help students understand and justify their reasoning about rates, we do believe that the instructional sequence we describe here offers a novel and

The analysis reported in this paper was supported in part by the National Science Foundation under grant No. REC-9619102. The opinions expressed do not necessarily reflect the views of the Foundation.

yet classroom-viable approach for helping students reason about rate in the context of measuring and comparing the speeds of different moving objects.

THEORIES ON THE TEACHING OF RATES AS MEASURES

Several different theories describe how students learn rational-number concepts (cf. Behr et al. 1992; Lobato and Thanheiser 1999; Thompson and Thompson 1994). As Lobato and Thanheiser (1999) note, most of these approaches can be characterized by either (*a*) the *representational* perspective, which focuses on creating drawings and learning conventional representations, such as graphs and algebraic notation, or (*b*) the *quantitative* perspective, which focuses on learning about how proportional quantities are related. In brief, the representational perspective emphasizes using a variety of mathematical representations to help students draw mathematical relationships. For example, when arguing for a focus on students' symbolizations, Behr and his colleagues describe how children can learn about proportional quantities by developing drawings to represent ratios, such as four oranges in each bag or three cookies for eight people.

In contrast to the focus described above on drawings and conventional representations such as graphs proposed by proponents of the representational perspective, Lobato and Thanheiser (1999) state that the primary focus of the quantitative perspective is "the sense one makes of the quantities in situations to which one applies numbers and calculations" (p. 291). One emphasis of the quantitative approach that bears particular relevance to the teaching of rates is the distinction between *extensive* and *intensive* quantities. Extensive quantities can be directly measured in single-dimension units. Examples of extensive quantities include the number of meters traveled or the mass of a rock. In contrast, intensive quantities are measured by considering the ratio of two extensive quantities. Examples of intensive quantities include speed (the ratio of the number of meters traveled *per* unit of time) and density (the ratio of a rock's mass *per* unit of volume). One of the major assumptions of the quantitative approach is that when studying speed, students should first identify the two extensive quantities (distance and time traveled) and then describe the way in which speed quantifies the intensive relationship between them.

The work of Thompson and Thompson (1994) provides one example of a curriculum that encompasses this distinction. Students' conceptions of speed can be seen as developing along the following continuum: (1) speed as a quantification of motion; (2) speed as completed motion measured in "speed lengths" (segments of distance traveled for a given unit of time); (3) speed as a *ratio* of completed distance to completed time; and (4) speed as a *rate* (direct proportional relationship) between any fractional amount of distance traveled and corresponding fraction of time elapsed.

The instructional approach that we took drew on both the quantitative and representational approaches so that we could accomplish two goals. First, we wanted students to be aware that the two quantities that compose speed are distance and time traveled. Although this might seem obvious to adults, research has indicated that students' prior experiences with speed (e.g., riding in a car and measuring speed with a speedometer) often cause them to gloss over or be less aware of the contribution made by the two extensive quantities of distance and time. Our second goal was for students to construct their own understandings of the intensive relationship between these two quantities. To help build such an understanding, we drew on the representational approach to focus students' attention on how position and velocity graphs represent motion. The representations the students used were dynamic graphs included in a software program called SimCalc, which links the motion of animated characters to distance and velocity graphs. The beauty of this program is that students can manipulate the graphs in order to ask "what if" questions and look at a motion simulation that provides visual feedback.

SimCalc Software

The SimCalc MathWorlds program was designed by mathematics educators at the University of Massachusetts Dartmouth. This program, which is available free of charge on the Internet at www.simcalc.umassd.edu, is a dynamic microworld for exploring one-dimensional motion. In this world, any combination of three graphs (position, velocity, and acceleration) can all be linked to an animated simulation and to one another (Kaput and Roschelle 1997). A student using the program can create or modify a position graph or a velocity graph and view the character's corresponding change in the simulation and on the other graphs. For example, figure 19.1 shows the linked velocity and position graphs for two characters, Frog and Clown. Using this world, a student can ask a question such as "What if I want to make Frog go faster?" She can investigate this question by, for example, changing Frog's velocity graph and then noticing how this change is reflected on the corresponding position graph and also in the simulation. One question that we focused on as we created the sequence was, how do we develop activities that would encourage students to ask "what if" questions so that they can explore speed by manipulating the relation between distance and time? We address this question by describing the ways in which students explained how they measured and compared several different characters' speeds.

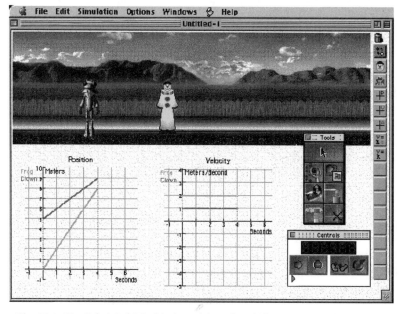

Fig. 19.1. SimCalc MathWorlds (1996) interface (Classic version for Macintosh) showing animated walking world with Frog and Clown, linked position and velocity graphs, world ruler, tools palette, and time controls palette

THE CLASSROOM SETTING

The lessons described in this article were part of a three-week instructional sequence that was conducted in a seventh-grade classroom in an urban middle school in Southern California. The students met in a Macintosh computer lab for seven of the fifteen class sessions. In addition, the teacher used a computer connected to an LCD projector to support some of the class discussions that were held in the regular classroom. The class contained fourteen girls and sixteen boys, all of whom were between the ages of twelve and fourteen. Two researchers sat in the back of the room during whole-class discussions and circulated around the computer lab to help the teacher field students' questions.

THE INSTRUCTIONAL SEQUENCE

In this article, we describe three of the instructional activities from the larger sequence. Readers who are interested in viewing all the worksheets are invited to download the activities from the digital database featured on the SimCalc Web site.

Activity 1: Interpreting Position Graphs of Completed Trips

The first few activities were designed to support students' efforts to understand position graphs by looking at the meanings of particular points, such as the starting and ending points of a trip. In this way, we hoped students would become aware of how the quantities of distance and time were both needed in order to describe the significant aspects of a completed trip. For example, in one activity students were given a description of a trip, such as "Frog started at 2 meters and walked to 10 meters in 4 seconds." After reading the information, they were asked to draw what a position graph of this trip would look like using pencil and paper. Then, they were given the opportunity to test their predictions by entering the graphs they had drawn into the computer and then running the linked simulation. If the resulting simulation showing Frog and Clown moving did not match the given description of the trip, the students were invited to manipulate the dynamic graphs until the simulation matched the initial description. As we predicted, the students' efforts to complete this task involved identifying the critical points of the trip. However, we did not predict that students would also make such observations as noting that a "steeper slope" meant the character traveled faster. These discussions not only enabled us to meet our first instructional goal of supporting students' efforts to interpret the two extensive quantities that compose the position graph but also enabled the teacher to build on students' observations about their intensive relationship.

Activity 2: Developing Notions of Speed as a Ratio

Once students had developed ways of explaining their graphical reasoning with specific correspondences between distance and time on a position graph, we shifted our focus to supporting their notions of speed as a ratio of these two quantities. For example, in one of the first whole-class discussions, the teacher asked the students to compare the speeds of two vehicles on the basis of their position graphs. In the following excerpt, the teacher used an overhead projector to show an animation featuring a red sports car and a blue van, each traveling toward the right of the screen.

Teacher(T): Let's look at this story. Now, someone has asked me how I made one car go faster than the other on the position graph. How would I tell someone which car went faster?

Marc: I have the answer. [You can] show the difference between their seconds. Show them by how many seconds it takes [each of the cars] to get to 60 [meters].

Rob: You can make it with slope. Have the sports car [i.e., the red graph] have a steeper slope.

T: What do you mean by slope?

Rob: The angle.

Cindy: A bigger hill.

T: [Speaking to the whole class] Why don't you each draw a graph for each of these cars on your worksheet?

The teacher then steps through the simulation second-by-second to show that the red sports car started at 0 meters and ended at 60 meters in 4 seconds, and the blue van started at 0 meters and ended at 60 meters in 6 seconds.

T: What did you do, Nate?

Nate: I put a dot at 4 seconds and up to 60. Then I put a line and connected it to (0,0). Then I put a dot at 6 seconds and a line to it [from (0,0)].

T: I'm curious; how fast do you think he was going?

Callie: He went 60 meters in 4 seconds.

T: Well, how fast is that? When you go on a ride at Magic Mountain [a local amusement park], you don't say, "I went 100 miles in 2 hours." What do you say? How do I describe "how fast"?

John: From the slope.

Ray: By seconds in meters. No, wait, meters in seconds.

Terry: You can tell because it [the red sports car] got to 60 meters faster than the van. Both are going the same meters, but one took less time.

T: But then how can I compare?

Nate: You go to 1 second and you see that it intersects at 15 meters, so 15 meters in one second.

T: Do others agree with that?

Laura: I do, because in the second second he [the red sports car] is at 30 meters, so that is 15 more meters.

T: Now look at the van; how fast is it going?

Kelly: 10 meters in one second.

T: Which is faster?

Students: 15 [meters in one second]!

One reason that we found this conversation compelling was that although it was the teacher's first attempt to explore the students' ideas about speed, we can see that the students had developed some ideas about the involvement of distance and time. More important, they mentioned their hypotheses regarding how the slope of the position graph appeared to be related to a car's speed. Regarding pedagogy, it is also interesting to note that instead of simply telling the students how to compute speed by measuring slope or calculating the ratio of distance and time, the teacher asked questions that were designed to guide the students' thinking about how the two quantities could be used to compare the cars' speeds. For example, instead of instructing the students to simply follow Nate's procedure for graphing, the teacher continued to ask different students to focus on explaining their ideas for measuring and comparing the speeds of the two cars. In so doing, the teacher was implicitly encouraging students to view their peers as resources for explanations and ideas.

Our explanation for how the students were able to develop these intuitions about speed is that they appeared to be *reasoning through* their prior experiences with manipulating the dynamic position graphs in the SimCalc Math-Worlds. In other words, we are claiming that their evolving interpretations of speed as a ratio between distance and time were supported by their experiences with changing the slope of the dynamic position graphs. This development can be seen in the progression of students' justifications that made references to their experiences with the graphs. For example, the discussion above began with a description of speed in regard to the finished journey. Next, the students began discussing the relevance of comparing the slopes of the two graphs. Following that, Nate described his method of determining speed by looking at how far the car has traveled after one second. After Nate described his method, we observed that many other students engaged in this practice during ensuing activities. As we walked around the lab, the students' explanations of their activity indicated that they were coming to define speed as a measure of the distance traveled in one second, which is a relatively sophisticated conception along Thompson and Thompson's continuum.

Activity 3: Viewing Position and Velocity Graphs as Controls for Motion

The culminating activity in this sequence had students work in pairs to plan, create, and present computer-based animations of their own design.

After the teacher showed as examples stories that featured different scenes and characters (included as demonstration files bundled with the SimCalc software), the students began designing their own stories to model. Their storyboards included such creative ideas as aliens hovering over a pond to kidnap a baby duck and sports cars breaking out of traffic jams. To ensure that students would not simply play around with graphs and then claim that the resulting story was their intention, we asked them first to complete a detailed description of the story they wanted to model. Examples of these planning sheets, which are shown in figures 19. 2 and 19. 3, also included predictions of the range of values that would be needed on the x- and y-axes for both the position and velocity graphs as indicated by the square boxes at the corner of each axis. After the students had completed their projects in the computer lab, each pair gave a short presentation to the class using the overhead LCD projector. Our claim regarding the importance of developing students' "what if" propensity is explored in the following descriptions of two students' experiences with moving the variables on the graphs in order to achieve their goals.

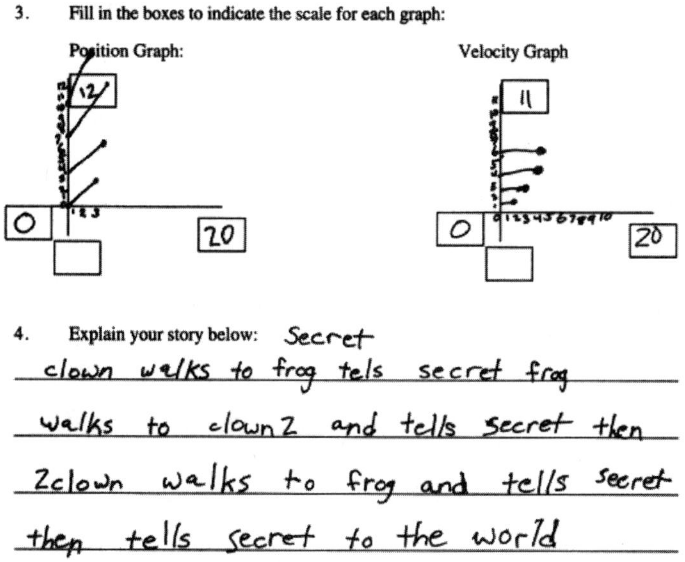

Fig. 19. 2. Charlotte's project description including sketches of position and velocity graphs

Charlotte's project, which is described in her planning sheet shown in figure 19. 2, involved having clowns and frogs tell each other a secret. She and her partner envisioned that there would be about three secret exchanges. At the end of her story, the secret would be revealed as an alien flew across the screen. The worksheet in figure 19. 2 shows how she and her partner envisioned the graphs would look. What is interesting is that when designing the project planning sheet, we had intended that the students would fill in the minimum and maximum ranges for the axes of each graph (by writing numbers in the small boxes), but we did not expect that students would actually draw their predictions of what the velocity and position graph plots would be. As shown in figure 19. 2, however, Charlotte did fill in the graph plots as well as some of the axis-dimension boxes. From her predictions, it is clear that she had not anticipated the problem of making each character wait in place until it was its time to move. Therefore, after Charlotte and her partner showed and explained their successful project to the whole class, the teacher asked Charlotte how they were able to get each of the characters to "wait" its turn to hear the secret.

> *Charlotte:* We took the first person to go, and we made him start walking. We made all the other people have a straight line [*points to the position graph*] so they would stand there until the character got there and then waited a couple of seconds. Then the next character would start walking.

From her explanation, it appears that Charlotte's group used a "straight line" on the position graph to control the characters' waiting times. This indicates that her group, which was representative of the majority, conceptualized velocity as a change of zero in distance over a given period of time.

In contrast to Charlotte's approach to controlling velocity, Anna reported that her group used the velocity graph to control their characters' speeds. What is particularly interesting about this approach is that Anna's story description (shown in fig. 19. 3) is focused exclusively on position and time markers. Yet, when they wanted to control speeds, they chose to use the velocity graph.

> *Teacher:* Was there anything that you thought was really difficult that you were glad you figured out?
>
> *Anna:* Yeah, we wanted the [Volkswagen] beetle [indicated with a green graph] to go faster than the duck 'cause we wanted the beetle [i.e., the green Volkswagen car] to win. We got the duck to go faster than the VW beetle, and it was difficult to get it the other way around.
>
> *Teacher:* How did you figure it out?
>
> *Anna:* We saw that if we moved the velocity like this [*points at the velocity plot for the VW bug and indicates an upward motion with her hand*], it went faster.

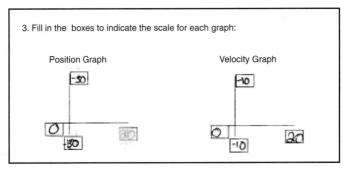

Our story is called flash in our story a duck and a car race. The car is [a VW beetle] starts at 10. The
duck started at 10. Then the car started racing. Started very slowly at 0 seconds. Well, the duck went
very fast and then they right passed [sic] each other and the duck loses and the beetle (VW car] wins
and our story is 20 seconds log [sic]

Fig. 19. 3. Anna's detailed explanation of a story project for MathWorlds

Anna's explanation indicates that her group developed a view of the velocity graph by asking, "What if we move the [horizontal] velocity segment?
Will that make the VW go faster?" The result was that her group came to
view velocity as a quantity that was controlled by both the velocity graph
(directly) and the position graph (indirectly through the slope). More
important, in an interview that took place after the sequence had been
enacted, Anna's answers to several questions indicated that she had begun to
view speed as a ratio measure of distance traveled per unit of time and hence
as a useful tool for making one character go faster than another.

In summary, we found that the students had progressed during the three
weeks to viewing speed as a ratio of distance and time traveled and, for
some, as a rate measuring distance traveled per unit of time. For example,
we believe that students such as those in Anna's group, who were reasoning
about speed as a rate, were able to make sense of the velocity graph. However, we do not claim that all the students had progressed to a the level of conceiving of speed as a rate—a level that would be consistent with activities
Thompson and Thompson's fourth level, which involve multiplicatively
comparing distance traveled and the amount of time it takes to go *any portion of that distance*. In fact, we have found in other research with a high
school Algebra 1 class that this final transition requires not only a firm grasp
of concepts of speed but also a facility in considering fractional distances
and corresponding fractional speeds, which these seventh graders had not
yet developed. What we do claim is that their experiences with the dynamic
graphs provided a strong experiential base for supporting further reasoning
about speed as a direct proportional relationship.

CONCLUSION

There are two implications from our work that may be useful to other teachers considering using interactive software like SimCalc to teach concepts of proportional thinking. First, most students do not come to mathematics classes with a natural "What if" propensity or a need to explain the mathematics behind the results they see on a computer screen. However, when given the opportunity to "play" with dynamic graphs in order to achieve a goal, the students in our class did develop hypotheses regarding the ways in which different manipulations (such as changing the length of time or the distance traveled) affected the slope of the graph and how this slope related to speed. However, we do not claim that this happened as a result of the computer alone. In fact, we believe that our strong emphasis on identifying the explicit quantities on the position graph, which was supported by the main tenet of the quantitative perspective, was a crucial first step in supporting students' efforts to understand the intensive relationship involved in understanding speed as a ratio and, ultimately, speed as a rate.

A second implication, which relates to the NAEP analysts' call for more emphasis on students justifications, is that the students' arguments improved most dramatically once they shifted from viewing the computer and the teacher as sole arbiters of correct answers to viewing their classmates' arguments as resources for explanations about rates. In fact, when the students began to develop their own arguments for why certain graphs would produce certain outcomes, they developed a deeper appreciation of the mathematical relation of speed as a ratio of distance and time. We believe that one necessary condition for realizing these implications was the teacher's efforts to help the classroom community develop a language and orientation for explaining not only how but also *why* these manipulations (such as changing the slope) had the effects on the speeds that they did.

REFERENCES

Behr, Merlyn, Guershon Harel, Thomas Post, and Richard Lesh. "Rational Number, Ratio, and Proportion." In *Handbook of Research on Mathematics Teaching and Learning*, edited by Douglas A. Grouws, pp. 296–333. New York: Macmillan, 1992.

Kaput, James, and Jeremy Roschelle. "Deepening the Impact of Technology beyond Assistance with Traditional Formalism in Order to Democratize Access to Ideas Underlying Calculus." In *Proceedings of the Twenty-first Conference of the International Group for the Psychology of Mathematics Education*, Vol. 1, edited by Erik Pehkonen, pp. 105–12. Lahti, Finland: University of Helsinki, 1997.

Kouba, Vicky L., Judith S. Zawojewski, and Marilyn E. Strutchens. "What Do Students Know about Numbers and Operations?" In *Results from the Sixth Mathematics Assessment of the National Assessment of Educational Progress*, edited by

Patricia A. Kenney and Edward A. Silver, pp. 87–140. Reston, Va.: National Council of Teachers of Mathematics, 1997.

Lobato, Joanne, and Eva Thanheiser. "Rethinking Slope from Quantitative and Phenomenological Perspectives." In *Proceedings of the Twenty-first Annual Meeting of the North American Chapter of the International Group for the Psychology of Mathematics Education,* Vol. 1, edited by Fernando Hitt and Manuel Santos, pp. 291–97. Columbus, Ohio: ERIC Clearinghouse for Sciences, Mathematics, and Environmental Education, 1999.

MathWorlds, Version 1.1b4. SimCalc Project, University of Massachusetts Dartmouth, 1996. www.simcalc.umassd.edu

Thompson, Patrick W., and Alba G. Thompson. "Talking about Rates Conceptually, Part I: A Teacher's Struggle." *Journal for Research in Mathematics Education* 25 (April 1994): 279–303.

Classroom Challenge

Clair Haberman

Mathematics Teacher, Saint Paul, Minnesota

Number Patterns

Calculate and find a pattern:

$$\frac{1}{2} + \frac{1}{6}$$

$$\frac{1}{6} + \frac{1}{12}$$

$$\frac{1}{12} + \frac{1}{20}$$

$$\frac{1}{20} + \frac{1}{30}$$

$$\frac{1}{30} + \frac{1}{42}$$

$$\vdots$$

Hint: If the denominators are factored, the exercises become

$$\frac{1}{1 \cdot 2} + \frac{1}{2 \cdot 3}$$

$$\frac{1}{2 \cdot 3} + \frac{1}{3 \cdot 4}$$

$$\frac{1}{3 \cdot 4} + \frac{1}{4 \cdot 5}$$

$$\frac{1}{4 \cdot 5} + \frac{1}{5 \cdot 6}$$

$$\vdots$$

Calculate and find a pattern:

$$\frac{1}{3} + \frac{1}{6}$$

$$\frac{1}{4} + \frac{1}{12}$$

$$\frac{1}{5} + \frac{1}{20}$$

$$\frac{1}{6} + \frac{1}{30}$$

$$\frac{1}{7} + \frac{1}{42}$$

$$\vdots$$

Calculate and find a pattern:

$$\frac{1}{2} - \frac{1}{3}$$

$$\frac{1}{3} - \frac{1}{4}$$

$$\frac{1}{4} - \frac{1}{5}$$

$$\frac{1}{5} - \frac{1}{6}$$

$$\frac{1}{6} - \frac{1}{7}$$

$$\vdots$$

Classroom Challenge

Susan Nickerson

San Diego State University, San Diego, California

Candy Bars and Lawns

Pretend that the following rectangles are candy bars. Cut each one into pieces to make the statements below each rectangle true.

1.

Sammy's part is 1/2 as large as Felicia's part.

Felicia's part is _____ times as large as Sammy's part.

How much of the candy bar is Felicia's?

What is the ratio of Sammy's part to Felicia's part?

2.

Huyen's part is 4/5 as large as Shana's part.

Shana's part is _____ as large as Huyen's.

How much of the candy bar is Huyen's?

What is the ratio of Huyen's part to Shana's part?

3. This time the rectangle represents a lawn that is mowed by Delia and Silvia. Delia arrived earlier and began before Silvia. The part she mowed is 3/2 as large as the part Silvia mowed.

Cut the lawn to represent Delia's and Silvia's parts.

The part Silvia mowed is _____ as large as the part Delia mowed.

They were given $25 for mowing the lawn. What would be a fair way to share the money? Explain your answer.

Classroom Challenge

Frank Lester

Indiana University, Bloomington, Indiana

Condo Problem

In an adult condominium complex, 2/3 of the men are married to 3/5 of the women. What part of the residents are married?

Possible Solutions:

1. Draw a picture. Keep drawing sets of women and men until you reach a point where the number of married men and women are equal. [Circles indicate married individuals.]

	Married
ⓂⓂ M ⓌⓌⓌ W W	2 men and 3 women
ⓂⓂ M	4 men and 3 women
ⓌⓌⓌ W W	4 men and 6 women
ⓂⓂ M	6 men and 6 women

Total of 9 men and 10 women ——→19 people
6 married men and 6 married women →12 people

Thus, 12/19 of the residents are married.

2. Recognize that you need to keep to the ratios given, but you need equal numbers of married men and women. So you need common numerators for 2/3 and 3/5. Since the least common multiple of the numerators (2 and 3) is 6, rewrite the fractions.

2/3 = 6/9 and 3/5 = 6/10.

Thus, you have 6 married men and 6 married women (12 married people out of 9 men total and 10 women total (19 people). So 12/19 of the residents are married.

3. Write equations. Then pick an appropriate number to generate a solution.

$2/3\ m = 3/5\ w \longrightarrow m = 9/10\ w$

Let $w = 10$, then $m = 9 \longrightarrow$ total number of people is 19

$2/3\ (9) = 6$ married men and $3/5\ (10) = 6$ married women $\longrightarrow$ 12 married people

Thus, 12/19 of the residents are married.

4. Guess and check. Pick a number for the number of married couples. [Choose one that is "nice" — that has enough factors that it is likely to work out.]

Suppose 60 couples are married. then $2/3\ m = 60$ so $m = 90$. $3/5\ w = 60$ so $w = 100$. The total number of people is 190 and the total number of married persons is 120. Thus, 120/190 or 12/19 of residents are married.

5. Draw a picture. The shaded area in the first figure represents the number of married men. But it is also the number of married women. Thus the shaded portion is 3/5 of the total number of women, as shown by the shaded portion and an additional 2/5 of the shaded portion shown in the second figure. The third figure shows that 6 of every 9 men are married and that 6 of every 10 women are married. The ratio of men to women in the town is 9 to 10.

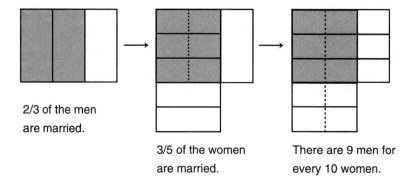

2/3 of the men
are married.

3/5 of the women
are married.

There are 9 men for
every 10 women.

20

Professional Development Introduction

David Duncan
for the Editorial Panel

PROPORTIONAL reasoning, including its associated concepts, operations, and applications, presents difficulties for some students. A good part of this difficulty involves the correct use of language. In everyday language and in the popular press, some mathematical phrasings are misused. Phrases like "200 percent larger than $1.2 million" and "200 percent as large as $1.2 million" are often misused as meaning the same, although in the first case, the result is $3.6 million, and in the second case it is $2.4 million.

Teachers should not only include attention to addition and multiplicative comparisons, but they should also give attention to the differences in phrasing, as suggested in these cases. An easier comparison to make in the classroom might be to think about the problem "My allowance is $10; yours is $10 more than mine." As shown in the first case in figure 20.1 below, this comparison is purely additive. The second case in the diagram shows "My allowance is $10; yours is 10 percent more than mine." Here the comparison is a multiplicative one, but the percent is then added to the original $10.

Fig. 20.1

In both the preceding example and in a multitude of other situations, teachers must have a diverse repertoire of approaches they can use to assist students in developing deep understanding of this type of reasoning. To use these approaches effectively, teachers should be familiar with not only the classroom techniques themselves, but also the background, both pedagogical and mathematical, that undergirds them. The purpose of this professional development section is to discuss many aspects of this background, together with their classroom implications.

We hope that all teachers will read this section reflectively, discuss it with colleagues, and use the ideas in it to guide their classroom practice.

21

Developing Students' Proportional Reasoning: A Chinese Perspective

Jinfa Cai

Wei Sun

PROPORTIONAL reasoning is one of the important forms of mathematical reasoning. The development of students' proportional reasoning can be regarded as the gateway to success in studying algebra. Proportional reasoning involves "a sense of covariation, multiple comparisons, and the ability to mentally store and process several pieces of information." (Post, Behr, and Lesh 1988, pp. 79) Mathematically, a proportional relationship can be represented by the function $y = mx$ (direct proportionality) or $xy = m$ (inverse proportionality), where m is a constant. Proportional relationships provide a powerful means for students to develop algebraic thinking and function sense. Psychologically, proportional reasoning is a way of thinking that involves making sense of quantitative relationships and comparing ratios. Proportional reasoning can be used to solve a variety of real-life problems.

Although the importance of developing students' proportional reasoning in school mathematics curriculum is recognized internationally, topics related to ratio and proportion are formally introduced at different grade levels in different countries. For example, in the United States topics related to ratio and proportion are formally introduced in the middle grades and revisited throughout the secondary school curriculum as students encounter topics such as similar figures and functions. In contrast, in China, the topics

Preparation of this manuscript is supported, in part, by grants from Spencer Foundation. Any opinions expressed herein are those of the authors and do not necessarily represent the views of Spencer Foundation.

related to ratio and proportion are formally introduced in elementary school (State Education Commission 1992).

It is instructive to examine how the Chinese elementary school mathematics curriculum introduces the concepts of ratio and proportion and other related concepts. Note that the focus is neither on the evaluation of the Chinese elementary school mathematics curriculum, nor on the comparison of Chinese and U.S. curricula. Instead, the focus is to provide a Chinese perspective on developing students' ability to reason proportionally. It is hoped that this international perspective will increase U.S. teachers' experience and awareness when they strive to help students make sense of proportional relationships.

In the Chinese elementary school mathematics curriculum, the topics related to ratio and proportion are divided into three units: ratio, proportion, and the application of ratio and proportion. The ratio unit is included in the first half of the fifth-grade curriculum (Division of Elementary Mathematics 1995a). The next two units are included in the second half of the fifth-grade curriculum (Division of Elementary Mathematics 1995b). The teachers' reference books recommend five lessons (40–45 minutes each) for teaching the concept of ratio, eight lessons for teaching the concept of proportion, and nine lessons for teaching the applications of ratio and proportion (Division of Elementary Mathematics 1995a, 1996b). The percent unit is placed after the ratio unit, but before the proportion unit and the application unit, which includes various problems to solve. The percent unit is introduced as a special type of ratio. In fact, *percent* in Chinese literally means "percent ratio." (Division of Elementary Mathematics 1995b)

TEACHING THE CONCEPT OF RATIO

In the Chinese elementary school mathematics curriculum, ratio is introduced after division with fractions. The concept of ratio is clearly defined as the comparison of two quantities with multiplicative relationships. Interestingly, *ratio* in Chinese means "to compare or comparison." The teachers' reference book highlights both the importance of ratio and difficulties students have understanding the multiplicative nature of the concept of ratio. Ratio is introduced and treated both as a concept and as an operation. When it is introduced, rather than just telling students what a ratio is and how it is represented, division is used as a bridge to connect the concept of ratio and its representations. The following example illustrates how Chinese teachers usually introduce the concept of ratio.

Example 1: Miller Middle School has 16 sixth-grade students, and 12 of them said that they are basketball fans. The remaining students are not basketball fans. How could we describe the relationship between the students who are basketball fans and those who are not?

Because there are 16 students and 12 of them are basketball fans, the number of students who are not basketball fans is $16 - 12 = 4$. By comparing these two numbers, some students may say that there are 8 more students who are basketball fans than those who are not basketball fans ($12 - 4 = 8$). Other students may say, "There are three times as many basketball fans as there are students who are not basketball fans ($12 \div 4 = 3$)." Yet others may say, "Among these 16 students, for every three students who like basketball, there is one student who does not like basketball." Students in the first case used additive reasoning to describe the relationship between the students who are basketball fans and those who are not. Students in the latter cases used multiplicative reasoning to describe the relationship between the students who are basketball fans and the students who are not. The multiplicative relationship between two quantities is defined as a ratio in Chinese textbooks.

There are two ways to describe how two quantities are related: additive or multiplicative. The ratio of the two quantities involves the comparison of the quantities using multiplicative relationships. The Chinese teachers' reference book indicates that such multiplicative relationships are considered as important, but difficult for some students to learn (Division of Elementary Mathematics 1996a). Deriving the concept of ratio by examples such as these has three advantages, which are listed below.

(1) Through such examples, teachers can help students distinguish multiplicative relationships from additive relationships.

(2) These examples start with an operation that students are familiar with; thus, they can easily see that a ratio can be considered as an operation.

(3) Students can also see how the ratio $a{:}b$, the division $a \div b$, and the fraction a/b are related.

In the example above, the Chinese mathematics teachers emphasize that when using a ratio to describe the relationship between the two groups of students, students need to begin with the knowledge they have learned (the division $12 \div 4$) and use this knowledge to describe the new situation (ratio of $3{:}1$). Because students have studied the division of integers and understand that a fraction also represents division, the connections among ratios, division, and fractions are easily made. The division $a \div b$ is the operation of the ratio of $a{:}b$, and fraction a/b is the result of the ratio operation. Chinese teachers call a/b the *value of the ratio* of $a{:}b$. By using the term the *value of the ratio*, which is represented by the fraction a/b, students are able to make connections among $a{:}b$, $a \div b$, and a/b. Figure 21.1 shows the way the Chinese mathematics teachers use to describe the relationship between the two groups of students.

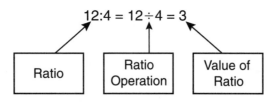

Fig. 21.1

Chinese teachers would explain that the ratio of the students who like basketball to the students who do not like basketball is 12:4 and the value of this ratio is 3. Also, a property on the value of ratio is introduced in the Chinese curriculum: when the two quantities in a ratio are multiplied by the same nonzero number, the value of the ratio remains the same. The property is introduced by relating it to reducing fractions. This property is very helpful for students when making comparisons between different ratios and for studying proportions (Division of Elementary Mathematics 1996a, 1996b).

TEACHING THE CONCEPT OF PROPORTION

The concept of proportion is built on the understanding of ratio. Chinese mathematics teachers have given a great deal of attention to connecting these concepts smoothly. Currently, a scale drawing is used as a bridge to connect ratio and proportion. Chinese mathematics teachers believe that if students have a solid understanding of scale drawing, they can quickly grasp the sense of proportionality. The following is an example of how a scale drawing helps students develop a sense of proportion.

Example 2: A corporation is going to build an office building. Two centimeters on the blueprint represents 5 meters on the ground. If the building is 18 centimeters long on the blueprint, what is the actual length of the building on the ground?

To solve this problem, students need to find the ratio of the distance on the blueprint to that on the ground. Chinese teachers emphasize the importance of using the same measurement unit for the two quantities to find the ratio. Therefore, five meters should be converted into 500 centimeters. In the context of scale drawing, ratios are usually reduced to unit ratios.

2 cm : 5 m   = 2 cm : 500 cm
            = 2:500
            = 1:250 (or 1/250)

Therefore, the ratio of the distance on the blueprint to that on the ground is 1:250. Initially, this sounds like a ratio problem, but while solving this problem, students are developing an understanding of proportion.

Let's take another look at this. The initial ratio is a ratio of two concrete quantities. By simplifying the ratio, it becomes an abstraction. This process from concrete to abstraction reflects the relationship between these two ratios. When teaching scale drawing, the Chinese mathematics curriculum emphasizes the meaning of scale and the role it plays in helping students determine proportion. With a good understanding of scale drawing, students will be able to make a smooth transition from ratio to proportion.

Proportion, in the Chinese mathematics curriculum, is described as two ratios that have the same value (Division of Elementary Mathematics 1995b). But students can easily get confused if they are not given the opportunity to see what this means. Thus, problems beyond scale drawing are used. The following example demonstrates how a Chinese teacher would introduce the concept of proportion.

Example 3: Car A traveled 120 miles in 2 hours. Car B traveled 180 miles in three hours. Describe which car traveled faster.

For Car A, the ratio of distance to time is 120:2 and for Car B, the ratio of distance to time is 180:3. They have the same value of ratio, 60 miles per hour, which is the speed that each car travelled. That is, $120/2 = 180/3 = 60$. Thus the concept of proportion can be naturally developed: two ratios that are equal.

In order to help students develop proportional reasoning, they need to understand the important role that direct proportionality and inverse proportionality play. Both direct proportionality and inverse proportionality represent a relationship between two quantities, and the change of one quantity directly affects the change of another quantity. In Chinese schools, students are given various examples to distinguish direct proportionality from inverse proportionality between quantities. They are also asked to make up problems in which quantities have direct proportionality and inverse proportionality (Division of Elementary Mathematics 1995b).

It is very important that students have concrete examples so that they can develop the understanding of direct proportionality through experience. However, many Chinese teachers believe that teaching proportionality cannot stop at the concrete level. A mathematical representation of direct proportionality needs to be generalized. Using the problem above, once students see how time and distance are related, it is natural to generalize the relationships among speed, distance, and time, which is $s = d/t$. From this mathematical representation (formula), students can easily see that when s is a constant, d and t are directly proportional.

Inverse proportionality is introduced in the Chinese elementary school mathematics curriculum right after direct proportionality is taught. This gives students a good opportunity to understand proportionality from a broader point of view. Meanwhile, students can also see how direct proportionality and inverse proportionality are related and distinguished from each other. They both represent a relationship between two quantities, and the change of one quantity impacts the change of another quantity. The following is an example from a Chinese mathematics textbook:

Example 4: A person plans to travel by bike to another village that is 60 km away. Discuss how the speed of the bike and time needed to travel are related.

Students can create a table to represent this relationship (see table 21.1). When students examine the table, they can see that as the speed increases, the time needed to travel 60 km decreases. Furthermore, the product of the corresponding quantities (speed and time) is a constant that is the actual distance between the two villages. This relationship represents the inverse proportionality. When students understand the relationship between speed and time, they can generalize it to a mathematical expression: $d = s \times t$. This generalization, again, is very important for students to understand how inverse proportionality is represented mathematically.

Table 21.1

Speed (km/hour)	1	2	3	4	5	6...
Time	60	30	20	15	12	10...

It is important to teach direct proportionality and inverse proportionality by helping students determine when two quantities are directly or inversely proportional. When the value of the ratio of the two quantities is a constant, then the two quantities are directly proportional. When the product of the two quantities is a constant, then the two quantities are inversely proportional. With a solid understanding of ratio, proportion, direct proportionality, and inverse proportionality, students have a strong foundation to develop their ability in reasoning proportionally.

THE DEVELOPMENT OF PROPORTIONAL REASONING

The Chinese curriculum contains a special unit on using proportional reasoning to solve problems in various contexts. The major learning objective in this unit is to enhance students' understanding of the concepts of ratio and proportion by applying a proportional relationship to solve various problems (Division of Elementary Mathematics 1996b). On the basis of the previous discussion of the concepts of ratio and proportions, students are guided to make connections between direct proportionality and inverse proportionality, connections between what they have learned and what they are learning, and connections among different solution strategies. The Chinese curriculum includes problems with various levels of difficulty, and the teachers' reference book contains discussions of teaching each type of the problems. In solving these problems, students first need to decide which proportional relationships among the quantities are involved in the problems, and they are encouraged to use different knowledge or approaches. Let's take a look at the following two examples.

> *Example 5:* John traveled 180 miles in three hours. It took him five hours to go from Chinatown to Germantown at the same speed. What is the distance from Chinatown to Germantown?

> *Example 6:* It took John five hours to go from Chinatown to Germantown when he traveled 60 miles per hour. If John wanted to spend four hours going from Chinatown to Germantown, how fast should he travel per hour?

This type of problem involves quantitative relationships among the distance, time, and speed. Before solving them, students are asked to determine the quantity that is invariant and the two quantities that are covariant. In Example 5, the speed is invariant, and there is a direct proportional relationship between the time and distance. In Example 6, the distance from Chinatown to Germantown is an unknown, which is invariant. With the invariant distance from Chinatown to Germantown, there is an inverse proportional relationship between speed and time. That is, the faster John travels, the less time he takes. Comparing the processes of solving these two problems may enhance students' understanding of the direct and inverse proportional relationships. Through qualitative reasoning and emphasis on invariant quantity, we can help students make sense of the standard proportional algorithm.

The Chinese teachers' reference book clearly indicates that teachers should guide students to solve these types of problems using different methods. Students are guided first to use the "arithmetic approach," with which they are very familiar, to solve the problem, and then to solve it through setting up an equation based on the proportional relationship.

> *Arithmetic Approach for Example 5*: Since John traveled 180 miles in three hours, he traveled 60 miles per hour. It took him 5 hours from Chinatown to Germantown, so the distance from Chinatown to Germantown is $5 \times 60 = 300$ miles.

> *Setting Up an Equation Based on Proportional Relationship for Example 5*: Let x = the distance from Chinatown to Germantown. Since John travels in the same speed, $180/3$ = the speed = $x/5$. Therefore, we have $180/3 = x/5$. Solving the equation for x yields $x = 300$ miles.

Similarly, students can be asked to solve the Example 6 in two different ways: (1) $5 \times 60 = 300$ miles. $300 \div 4 = 75$ miles per hour. (2) Let x be the number of miles John traveled per hour when he wanted to complete the journey in four hours. $4x$ = the distance from Chinatown to Germantown = 5×60. Therefore, $4x = 5 \times 60$; then solving the equation $x = 75$.

The arithmetic approach holds the promise for developing students' proportional sense through qualitative reasoning (Post, Behr, and Lesh 1988). The equation approach based on the proportional relationship can facilitate students' understanding of the standard proportional algorithm, which is a useful and efficient tool for problem solving. In Example 5, the equation was set as $180/3 = x/5$ because $180/3$ = the speed and $x/5$ = the speed. In Example 6, the equation was set as $4x = 5 \times 60$ because $4x$ = the distance from Chinatown to Germantown and 5×60 = the distance from Chinatown to Germantown. In addition, through solving the problems in two different ways, students can make connections between various methods. Such comparison is beneficial for both reinforcing what students had learned and setting up stages for learning new ways to solve a problem. In particular, it will foster students' development of algebraic thinking (Cai 1998).

It is not uncommon for Chinese teachers to change knowns or unknowns to make connections among instructional tasks in a lesson. After teachers and students solve a problem, teachers can ask students to look at a related problem. We discussed these two problems together in this article, but in a classroom Chinese teachers usually present them as one problem with two variations. Students are guided to solve the two variation problems while attending to the various ways in which the new problem (Example 6) relates to the original problem (Example 5). Examples 5 and 6 can be solved by directly applying a proportional relationship. After solving this type of problem, Chinese teachers usually pose more complex problems, such as Examples 7 and 8 shown below, for students to solve. Again, students are encouraged to solve each problem in different ways.

Example 7: Emma is reading a storybook. If she plans to read 6 pages per day, she can complete her reading in 12 days. She wants to speed up her reading, so she reads two more pages each day than she had planned. How many days will she take to complete her reading of the book?

Solution 1 for Example 7: $6 \times 12 = 72$ pages for the book. Since Emma reads two more pages than she planned, she actually reads 8 pages per day. $72 \div 8 = 9$ (days).

Solution 2 for Example 7: Let x be the number of days Emma spends to complete her reading of the book. $x(6 + 2)$ = total number of pages for the book = 6×12. Therefore, $x(6 + 2) = 6 \times 12$, $x = 9$.

Example 8: Johnson School District bought some mathematics books. The district distributed these books to the Edison Middle School and Radnor Middle School using a 4 to 5 ratio. Edison Middle School received 200 books. How many books are there in total?

Solution 1 for Example 8: Since these books were distributed according to the 4 to 5 ratio, we can imagine that they were divided into nine equal parts. Edison Middle School got 4 out of the nine parts, which is 200 books. Therefore, each part represents 50 books, resulting from $200 \div 4 = 50$; $50 \times 9 = 450$. Therefore, there are 450 books in total.

Solution 2 for Example 8: Let   be the total number of books the district has. Since these books were distributed according to the 4 to 5 ratio, we can image that they were divided into nine equal parts. Edison Middle School got 4 out of the 9 parts; therefore, the ratio of the number of books Edison Middle School got to the total number of books should be $4:(4 + 5)$. Thus, we have $200/x = 4/(4+5)$. By solving the equation for x, the total number of books the district had is 450.

Solution 3 for Example 8: Let x be the number of books that Radnor Middle School got. Since the district distributed these books to the Edison Middle School and Radnor Middle School according to 4 to 5 ratio, the ratio of the number of books Edison Middle School got to the number of books Radnor Middle School got should be $4:5$. Therefore, $200/x = 4/5$. By solving the equation for x, the total number of books Radnor Middle School got is 250. Therefore, the total number of the books should be $200 + 250 = 450$ books.

Solution 4 for Example 8: The total number of books the school district has can be considered as a "whole." Since the district distributed these books to the Edison Middle School and Radnor Middle School according to a 4 to 5 ratio, Edison Middle School got 4/9 of the total books. Let be the total number of books the district has; then, $x \times 4/9 = 200$. By solving the equation for x, the total number of books the district had is 450.

The teachers' reference book contains a detailed analysis of each of the solutions. For Example 8, Solution 1 involves the ratio of whole numbers. Solutions 2 and 3 are based on proportional relationship. Solution 4 is based on the part-whole relationship of the number of books Edison Middle School got and the total number of the books the school district had. According to the teachers' reference book, discussion of these solutions to the problem can help students make coherent connections among ratio, proportion, fraction, and part-whole relationships.

SUMMARY

There is no doubt that ratio and proportion are among the most important topics in school mathematics. However, we are still searching for better ways to help students understand these concepts and develop their proportional reasoning skills. This Chinese perspective of the teaching and learning of ratio and proportion is designed to develop students' proportional reasoning ability. The examination of curriculum and instructional practice in other nations can provide a broader point of view on how this topic might be treated. We hope that such international perspective can increase U.S. teachers' awareness when they try to address the issues and challenges facing in students' learning of the topics related to ratio and proportion.

REFERENCES

Cai, Jinfa. "Research into Practice: Developing Algebraic Reasoning in the Elementary Grades." *Teaching Children Mathematics* 5 (December 1998): 225–29.

Division of Elementary Mathematics. *Mathematics: Elementary School Textbook (Number 9).* Beijing: People's Education Press, 1995a.

———. *Mathematics: Elementary School Textbook (Number 10).* Beijing: People's Education Press, 1995b.

———. *Mathematics: Teachers' Reference Book (Number 9).* Beijing: People's Education Press, 1996a.

———. *Mathematics: Teachers' Reference Book (Number 10).* Beijing: People's Education Press, 1996b.

Post, Thomas R., Merlyn J. Behr, Richard Lesh. "Proportionality and the Development of Prealgebra Understandings." In *The Ideas of Algebra, K–12,* 1988 Yearbook of the National Council of Teachers of Mathematics (NCTM), edited by Albert P. Shulte, pp. 78–90. Reston, Va.: NCTM, 1988.

State Education Commission. *Mathematics Teaching Syllabus for Nine-Year Compulsory Education.* Beijing: People's Education Press, 1992.

22

The Development of Rational Number Sense

Irene T. Miura

Jennifer M. Yamagishi

LANGUAGE SUPPORT FOR RATIONAL-NUMBER UNDERSTANDING

THE Third International Mathematics and Science Study (TIMSS 1996) reported cross-national differences in mathematics achievement favoring students from Asian-language-speaking countries, particularly Singapore, Korea, Japan, and Hong Kong. Teaching strategies were cited as an important factor in explaining these differences. However, the educational context for culturally mediated teaching methods should not be dismissed. Children come to school with an intuitive or socially acquired knowledge of number, which affects the teaching and learning of mathematical concepts during formal school instruction. An examination of that number sense, which provides a context for various teaching strategies, is also important.

There has been renewed interest among researchers in the role that cultural processes may play in the understanding of mathematics and in the performance of mathematics tasks (Cobb 1993; Kaput 1991; Saxe 1988). Central to these processes is the child's acquisition of cultural tools for thinking and learning—one of the most important being language (Rogoff 1990; Steffe, Cobb, and von Glasersfeld 1988). The language of mathematics provides a cultural context for mathematical activities. Nunes (1992) has suggested that the use of culturally developed symbol systems restructures mental activity without altering basic abilities such as memory and logical reasoning. In other words, characteristics of languages (e.g., the structure of numeration systems that makes counting simpler in some languages than others) may make tasks easier to perform. Certain characteristics of Asian number languages (culturally developed tools for mathematics) may promote a developmental head start and affect the later performance of mathematical tasks. Comparative studies have suggested that language characteristics can influence cross-national differences in abstract counting performance (Miller et al. 1995) and the understanding of base-ten concepts (Miura et al. 1993; Song and Ginsburg 1987).

Language characteristics may also influence the understanding of numerical fractions. In a study conducted with first and second graders in Croatia, Korea, and the United States (Miura et al. 1999), children were asked to link numerical fraction symbols with their appropriate regional representations. The fractions used in the study were 1/3, 2/3, 2/4 (used twice), 3/4, 2/5, 3/5, and 4/5. Korean-speaking first and second graders were better able to identify the corresponding pictorial representation than were Croatian- and English-speaking children. By the beginning of grade two, 25 of the children in Korea (76 percent of all participants) correctly identified all eight fractions on the test. No child gave more than two incorrect responses. Only one child in the United States and no child in Croatia was able to identify all eight fractions correctly.

In Korean (as well as Chinese and Japanese), the concept of fractional parts is embedded in the mathematics terms used for fractions (e.g., *one-fourth* is spoken in Korean as "of four parts, one"). The oral term expresses the part-whole relation and may influence early conceptualizations of fractions. The spoken language connects the mathematical meaning of fractions, that is, the language matches the symbolic representation to verbal knowledge, which is a necessary step in the understanding of fractions (Mack 1993). In U.S. textbooks, the English term *one-fourth*, and the symbolic, 1/4, must be explained to children as meaning "one of the four equal parts." This is also the case in Croatia. The Korean vocabulary of fractions appeared to influence conceptual understanding and resulted in the children having acquired a rudimentary understanding of fraction concepts prior to formal instruction (Miura et al. 1999).

Thus, language characteristics may support children's intuitive understanding of the part-whole relation and a recognition of fractions as parts of a region. The effects on other interpretations of fractions, as a set subdivided into equal-sized parts, a ratio, an indicated division, or as an expression of rational numbers, were not examined in the study.

GRADE PLACEMENT OF FRACTION TOPICS

An examination of textbooks used by the participating schools in Croatia, Korea, and the United States shows that the introduction of fraction concepts varies by country and textbook publisher. The textbook used in the participating U.S. classrooms, *Invitation to Mathematics* (Scott, Foresman Co. 1988), introduces the concept of equal parts at the end of the first-grade text (pp. 243–45). The first-grade teachers did not complete these pages as part of the first-grade curriculum. In the second-grade textbook, again at the end of the text, the concept of fractions is presented as a part-whole problem. In the Croatian textbook, *Matematika 2* (Skolska Knjiga 1997), fractions are introduced at the end of grade two as a problem of division—the whole being divided into *x* equal parts. In Korea, fraction concepts are introduced in the first half of the second grade in Unit 8, which begins on page 82. The Korean textbook, *San Soo 2-1* (Ministry of Education 1997), introduces the concept first as equal parts of a whole, then continues with fractions as part-whole relations. The ten pages are limited to problems using geometric figures. In the United States, teachers may not get to those final pages in the text, and unlike Japan and Korea, a mastery of fraction concepts in grades one and two is not expected.

AN EXPERIMENT IN TEACHING

In *MathLand: Journeys through Mathematics, Grade 2* (Creative Publications 1995), the language of fractions is a five-day, one-week topic in Unit 8 of the grade 2 textbook. Fractions in the formal sense are not mentioned, and the focus is on experiences that will "form a basis for the understanding of fractions in the years to come" (Charles et al. 1995, p. 286). In the second year of using the Creative Publications textbook, Mrs. Y., a second-grade teacher, decided to augment the lessons by using corresponding fraction symbols and verbal expressions. If inherent language characteristics can support earlier concept acquisition, she wondered if direct language support might also enhance concept understanding.

Prior to beginning the lessons on fractions, Mrs. Y. assessed fraction understanding using the questionnaire taken from Miura et al. (1999). Each child was given a test sheet with four numerical fractions printed at the top. Mrs. Y. explained that the terms at the top of the page were called fractions. She pointed to the 1/2 and explained that this fraction was called one-half. The remaining fractions, 1/3, 1/4, and 1/5 were described in a similar fashion. Below the examples, there were eight numerical fractions at the left side of the page. These were fractions that were more complicated and that children would not ordinarily use (1/3, 2/3, 2/4, 3/4, 2/5, 3/5, and 4/5). Each numerical fraction was followed by four geometric figures (circles, squares, or rectangles) with varying portions shaded (pictorial representation choices), as shown in figure 22.1. The teacher explained that she would read the fraction at the left of the page and that the child was to draw a circle around the picture in the row that showed the fraction. Mrs. Y. did not prompt the children in any way.

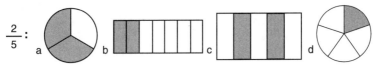

Fig. 22.1 Sample test item

The response choices were designed so that an error analysis could determine the strategy children might be using in defining fractions. In addition to random errors, the incorrect response choices included errors that would result if the child focused on the numerator only, the denominator only, or on both the numerator and denominator at the same time.

The results from this pretest showed that the children were unfamiliar with fraction symbols and their corresponding pictorial representations (table 22.1). For the 16 children, the mean number correct on the eight items was 0.81. An analysis of errors showed that a focus on both the numerator and the denominator (e.g., selecting 2/7—2 colored and 5 uncolored sections—for 2/5, as shown in fig. 22.1b) was the most common error made by the children. One child identified all eight fractions correctly, one made two correct responses, and three children made one correct choice. The remaining 11 children (69 percent of the participants) were unable to identify any of the eight items.

TEACHING STRATEGIES

Before beginning the first lesson, Mrs. Y. read *Gator Pie* by Louise Mathews, a story about two alligators who find a whole pie and want to eat it. The alligators discuss how they might cut the pie into two equal pieces, but before they can do so, a third alligator comes along. Mrs. Y. and the class discussed how to cut whole objects, the meaning of equality, and the idea that when using fractions to talk about parts of a whole, the parts must be equal. The mathematics text calls this sharing fairly between children; one-half of a pie must be exactly the same size as the other half of the given pie.

Fractions were explained as

$$\frac{\text{number of parts}}{\text{number of parts to make a whole (thing)}}.$$

For example, 1/4 means

$$\frac{1}{4} \quad \begin{array}{l} \text{part} \\ \text{parts to make a whole pie.} \end{array}$$

To record their answers, children were given precut shapes to use so that in addition to drawing a line through a picture cookie to share it with a friend, the student also cut a paper cookie into two equal parts. The cut pieces allowed children to see that the pieces are indeed equal. They were instructed to write the fractions below their answers: 1/2 + 1/2 = 2/2 = 1 whole cookie. Finally, the children were taught how to read the fractions aloud. This they did in unison, saying, "One-half plus one-half equals two halves equals one whole cookie."

There were five lessons: circles (cookies) divided with 1, 2, 3, 4, and 5 friends; liquid (tea) divided into cups between 2, 3, and 4 children; a square (sandwich) divided into rectangles for 2 and 3 people; a square (sandwich) divided for 4 and 5 people; a square (sandwich) divided into triangles for 2 people; and a square sandwich divided for 6.

A posttest using the same assessment questionnaire showed that children were better able to associate numerical fractions with their pictorial representations following instruction (table 22.1). For the 16 children for whom

Table 22.1
Number correct: pre- and postinstruction

Student	Pretest	Posttest
1	2	4
2	1	8
3	0	8
4	0	0
5	8	8
6	0	8
7	0	7
8	0	5
9	0	8
10	0	8
11	0	7
12	0	8
13	0	7
14	1	8
15	1	7
16	0	8

Note: The maximum possible was 8.

pre-and posttest scores were available, the mean number correct on the posttest for the eight items was 6.8. The range of correct responses was 0 to 8; 13 children made seven or eight correct responses, and only one child did not make any correct responses. A focus on both the numerator and the denominator continued to be the most common error made by the children.

CONCLUSIONS

In this experiment, direct instruction was given to help children see the connections among fraction symbols, verbal knowledge, and pictorial representations. Whereas in certain Asian languages, such as Korean, the language for fractions expresses the part-whole relation, in English it does not. Thus, in this experiment, the teacher explicitly explained the meaning of the verbal fraction and showed the children the pictorial representation to which the fraction might refer.

The fractions used in the lessons had denominators that were six or less. The numerator was always one, except when the fraction denoted a whole number. The posttest results suggest that children were able to extend their comprehension of the simple fractions to an understanding of more-complex fractions that they had not encountered, like 3/4 and 4/5.

Although the printed lessons in the language of fractions were described simply as a nonconceptual introduction to fractions, the lessons were easily broadened to include conceptual understanding as well. Language supports, when not inherently available, can be added to teaching strategies. We do this when we teach expanded notation in an effort to help children understand base-ten concepts, or introduce set theory when teaching multiplication. The teaching of fraction concepts, too, requires linguistic support, and an early introduction of that support may help develop rational-number sense from which to develop further rational-number understanding.

REFERENCES

Charles, Linda, Micaelia R. Brummett, Heather McDonald, and Joan Westley. *MathLand: Journeys through Mathematics, Guidebook, Grade 2*. Mountain View, Calif.: Creative Publications, 1995.

Cobb, Paul. "Cultural Tools and Mathematical Learning: A Case Study." *Journal for Research in Mathematics Education* 26 (July 1995): 362–85.

Creative Publications. *MathLand: Journeys through Mathematics, Skill Power, Grade 2*. Chicago: Creative Publications, 1998.

Kaput, James J. "Notation and Representation." In *Constructivism in Mathematics Education*, edited by Ernst von Glasersfeld, pp. 53–74. Dordrecht, Netherlands: Kluwer, 1991.

Mack, Nancy K. "Learning Rational Numbers with Understanding: The Case of Informal Knowledge." In *Rational Numbers: An Integration of Research*, edited by Thomas P. Carpenter, Elizabeth Fennema, and Thomas A. Romberg, pp. 85–106. Hillsdale, N.J.: Lawrence Erlbaum Associates, 1993.

Miller, Kevin F., Catherine M. Smith, Jianjun Zhu, and Houcan Zhang. "Preschool Origins of Cross-National Differences in Mathematical Competence: The Role of Number-Naming Systems." *Psychological Science* 6 (1995): 56–60.

Ministry of Education. *San Soo 2-1*. Seoul, Korea: Ministry of Education, 1997.

Miura, Irene T., Yukari Okamoto, Chungsoon C. Kim, Marcia Steere, and Michel Fayol. "First Graders' Cognitive Representation of Number and Understanding of Place Value: Cross-National Comparisons—France, Japan, Korea, Sweden, and the United States." *Journal of Educational Psychology* 85 (1993): 24–30.

Miura, Irene T., Yukari Okamoto, Vesna Vlahovic-Stetic, Chungsoon C. Kim, and Jong H. Han. "Language Supports for Children's Understanding of Numerical Fractions: Cross-National Comparisons." *Journal of Experimental Child Psychology* 74 (1999): 356–65.

Nunes, Terezinha. "Cognitive Invariants and Cultural Variation in Mathematical Concepts." *International Journal of Behavioral Development* 15 (1992): 433–53.

Rogoff, Barbara. *Apprenticeship in Thinking: Cognitive Development in Social Context*. New York: Oxford University Press, 1990.

Saxe, Geoffrey B. "Candy Selling and Math Learning." *Educational Researcher* 17 (1988): 14–21.

Scott, Foresman Co. *Invitation to Mathematics*. Glenview, Ill.: Scott, Foresman Co., 1988.

Skolska Knjiga. *Matematika 2*. Zagreb, Croatia: Skolska Knjiga, 1997.

Song, Myung-Ja, and Herbert P. Ginsburg. "The Development of Informal and Formal Mathematics Thinking in Korean and U.S. Children." *Child Development* 85 (1987): 1286–96.

Steffe, Leslie P., Paul Cobb, and Ernst von Glasersfeld. *Construction of Arithmetical Meanings and Strategies*. New York: Springer-Verlag, 1988.

Third International Mathematics and Science Study (*TIMSS*). *Mathematics Achievement in the Middle School Years*. Chestnut Hill, Mass.: TIMSS International Study Center, 1996.

23

Multiplicative Reasoning: Developing Students' Shared Meanings

Cristina Gómez

THE development of proportional reasoning is a difficult task. Although many people identify proportional reasoning with the use of the cross-multiplication method, research on students' understanding has shown the development of true proportional reasoning involves not only good sense of fractions and rational numbers but also proficiency in other areas such as ratio sense, relative thinking, partitioning, unitizing, and changing quantities (Lamon 1999). The development of proportional reasoning requires moving away from additive thinking to multiplicative ideas, particularly involving relative thinking.

Knowing how to participate in discourse practices is an important aspect of the ability to understand a specific content (Greeno, Collins, and Resnick 1996). The following classroom discussion shows how the students and the teacher participate in the process of "negotiating" the meaning of ratios and multiplicative relationships. Negotiating refers to the social interaction that takes place in this classroom and supports the development of students' personal identities and the construction of shared meanings (Wenger 1998).

Ms. V. teaches a combination fourth- and fifth-grade class in an elementary school in a mid-sized urban city in the Midwest. She had participated with other teachers and researchers in professional development programs designed to help teachers understand and build on children's mathematical

The data for this case study were collected and analyzed as part of a graduate seminar taught by Thomas Carpenter at the University of Wisconsin—Madison. Carrie Valentine, Celia Rousseau, Olof Steinthorsdottir, Lesley Wagner, Peter Wiles, Shih-Yi Chan, Gwen Fisher, and Christopher Hartmann collected the data and participated in the analysis.

Picture

16 people 20 eggs

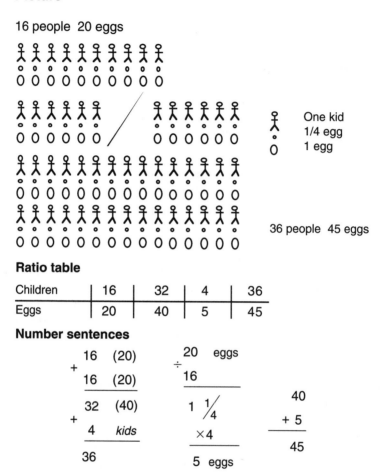

Ratio table

Children	16	32	4	36
Eggs	20	40	5	45

Number sentences

$$
\begin{array}{r}
16 \;\; (20) \\
+ \;\; 16 \;\; (20) \\
\hline
32 \;\; (40) \\
+ \;\; 4 \;\; kids \\
\hline
36
\end{array}
$$

$$
\begin{array}{r}
20 \;\; eggs \\
\div \;\; 16 \\
\hline
1 \; {}^{1}\!/_{4} \\
\times 4 \\
\hline
5 \;\; eggs
\end{array}
$$

$$
\begin{array}{r}
40 \\
+ \; 5 \\
\hline
45
\end{array}
$$

Problem: The White House held an Easter egg hunt on Monday. For every 16 children the staff hid 20 pink eggs. If there were 36 children, how many pink eggs did they hide?

Fig. 23.1

thinking (Carpenter, Fennema, and Franke 1997). Ms. V.'s class had been working with word problems that required proportional reasoning. During the two weeks previous to this episode, the students had been solving problems like the one presented in figure 23.1. The strategies used by these students showed their different levels of understanding and development of proportional reasoning. Pictures representing all the information from the problem, ratio tables using a doubling and adding strategy, and number sentences repeatedly using the doubling and adding strategy were the most common solutions the students presented in this class.

From their solutions, it is clear that the students had a good understanding of the situation and good sense of fractions. They recognized the quantities involved (number of kids and number of eggs) and the relative change between those quantities (for every kid there is 1 1/4 eggs, or, for every 4 kids there are 5 eggs). It is not clear, though, whether they saw the scaling factor within the quantities and the multiplicative relationships involved in the problem.

A "measure space" is a structure with four components: a set of objects, a quality of those objects, a unit, and a process to assign a numerical value to this quality. These structures are the foundation of all concepts related with measurement. In a direct proportion problem there are two measure spaces involved. In the problem given in Ms. V.'s classroom, the number of children and the number of eggs are the two measure spaces involved. True understanding of proportional reasoning involves the ability to see the two multiplicative relationships between the two measure spaces in the problem.

In the solution using the ratio table, the scaling factor changes for each entry in the table (e.g., between the first two entries the scaling factor is 2): the number of children and the number of eggs has been doubled. This scaling factor could be seen as a multiplicative relationship "within" each measure space. The other multiplicative relationship is given by the ratio between the number of children and the number of eggs (the number of eggs is 1 1/4 times the number of children). This ratio is then the multiplicative relationship "between" the measure spaces.

SETTING THE DISCUSSION

After solving some other problems involving different number relationships and different contexts, Ms. V. had decided to have a whole-group discussion. The purpose of the discussion was to help students to realize the different multiplicative relationships involved in ratio problems and to discuss efficient ways of solving these types of problem. The following vignette corresponds with the initial stage of the discussion.

Mike, a student, read the problem: Four tents will house 12 scouts. If there are 40 tents, how many scouts will have a place to sleep?

(1)   *Ms. V.:*   You know what I want to do today? I would like for you to tell me what to do to solve this problem. So, what should I do to solve this problem? Give

me an idea. I don't know how to do it. Peter, what should I do?

(2)   *Peter:*   Find out how many scouts go in one tent.

(3)   *Ms. V.:*   How many scouts go in one tent? Okay, so how do I do that? Peter?

(4)   *Peter:*   Well, four times three is twelve scouts. Divide twelve into four parts.

(5)   *Ms. V.:*   So, I could do twelve, and divide it into four parts, and that equals three, okay. [*Writes 12 ÷ 4 = 3*]

(6)   *Peter:*   And go three times forty.

(7)   *Ms. V.:*   Now, why do I do that?

(8)   *Peter:*   Because there are forty tents and three people in each tent.

(9)   *Ms. V.:*   Each, okay, say it again so that everybody can hear it.

(10)   *Peter:*   Three scouts and, wait, never mind. Three scouts in one tent.

(11)   *Ms. V.:*   Three scouts in one tent.

(12)   *Peter:*   It would be, well, if you do it forty times three, or you could do it three times forty.

(13)   *Ms. V.:*   Three times forty [*writes 40 × 3 = 120*]

(14)   *Peter:*   And then that would be 120.

(15)   *Maggy:*   I have another way to solve it, Ms. V.

(16)   *Ms. V.:*   What's that? You have another way to solve it? Can you hold on a second? Hold your thought. Karen, why did we go twelve divided by four equals three?

(17)   *Karen:*   To see how many people are in one tent.

(18)   *Ms. V.:*   Okay, and Tao, why did we do forty times three or three times forty makes 120?

(19)   *Tao:*   To get the answer.

(20)   *Ms. V.:*   Well, what does that information tell us? Why did we do this to get the answer? What does this stand for?

(21)   *Tao:*   One tent.

(22)   *Ms. V.:*   One tent, and what does the 40 stand for?

(23)   *Tao:*   Um, the tents.

(24)   *Ms. V.:*   The tents, okay. What does this [points at 3] represent?

(25)   *Tao:*   Three scouts in a tent.

(26)   *Ms. V.:*   Three scouts, okay. So, three scouts in a tent times the 40 tents we have, and we get 120 scouts. Okay, good.

Two important features of this vignette are the social interaction and the discourse forms that took place. The roles the students and the teacher took are noteworthy. The teacher gave students opportunities to participate, present their ideas, and "work through the problem under discussion while simultaneously encouraging each of them to listen to and attend to the solution paths of the others, building on each other's thinking" (O'Connor and Michaels 1996). The conversation was directed by the teacher, but the students had time, and were expected, to explain their responses—not just shout answers.

The discourse forms valued in this classroom were explanations, ideas, and procedures. Not only were the answers important, the reasoning behind was as important, too. The "negotiation of meaning" took place because the teacher functioned as a model asking questions that address the important ideas related with the mathematical content (lines 7, 16, 18, 20, 24). Even though students gave solutions, the conversation remained focused on why the process works (lines 17 to 26).

Alternative Solutions

But solving the problem was not Ms. V.'s only goal. She wanted to help her students see the different multiplicative relationships involved in the problem. To do so, she acknowledged other solutions, confident that her students would find other ways of looking at the problem. Maggy proposed the following alternative solution.

(27)   *Ms. V.:*   You have a different way to solve it?

(28)   *Maggy:*   Um, well, I know for a fact that four times ten is 40.

(29)   *Ms. V.:*   Okay, why do we want to do that?

(30)   *Maggy:*   Because you need to figure out how many scouts are in 40 tents. You need to figure out what to times … what you should times the scouts by.

(31)   *Ms. V.:*   Okay, so you want me to say four times ten is equal to 40. What is that going to give you? [*Writes* $4 \times 10 = 40$]

(32)   *Maggy:*   You need to know what to times the scouts by. So, if you do twelve times ten, you can have your answer, and that's 120.

(33)   *Ms. V.:*   So you do twelve times 10, and that equals 120.

(34)   *Maggy:*   Because two times ten is twenty and ten times ten is 100.

(35)   *Lee:*   They're both really short ways. They're both two steps.

(36)   *Ms. V.:*   They're both really short ways.

At this point, the two multiplicative relationships—the ratio and the scaling factor—are exposed and the solution of the classroom looks like the following.

$12 \div 4 = 3$ $4 \times 3 = 12$

$40 \times 3 = 120$

$4 \times 10 = 40$

$12 \times 10 = 120$

In this solution the multiplicative relationship between the measure spaces is 3 (the number of scouts is 3 times the number of tents) and the relationship within the measure spaces is 10 (the number of scouts and the number of tents are both increased 10 times).

Again, the norms of the class are illustrated here. The teachers asks questions that direct the class to her goal (lines 29, 31, 36) and a student, Maggy in this case, explains her reasoning as well. The meaning of "efficient strategy" as negotiated by this group of students, corresponds with the number of steps involved in the solution (line 35). In previous classes the concept of efficiency had emerged but the use of other tools like the ratio table made comparison of different strategies difficult. The strategies involving multiplicative relationships were valued as efficient, short, and easy.

LABELING

Another important part of the goal Ms. V. had for this class was to help students advance in their understanding of ratio. The class had an initial discussion about what they knew about ratios, but that was just the initial stage in the process of negotiation of meaning. At this point Ms. V. wanted to use the units of each measure space as tools to think about the ratios involved in the problem. This process of reflection about the solution process helped students realize the connections necessary to understand situations involving proportional reasoning. The following exchange illustrates this point.

(37)   *Ms. V.:*   Okay, so this number [*points to 10*] says how many times what?

(38)   *Ethel:*   How many ... ten groups of four tents.

(39)   *Ms. V.:*   Ten groups of four tents. Great! What's this 3 groups of? [*Points to 4 × 3 = 12*]

(40)   *Ethel:*   Three scouts for one tent. I don't know if it's groups.

(41)   *Ms. V.:*   Okay, you don't know if it's groups. Linda, help us out here. Is this groups?

(42)   *Linda:*   It's people in a tent.

(43)   *Ms. V.:*   Okay, in this one [*points to 12 × 10 = 120*] you did ten groups. What does this number mean?

(44)   *Linda:*   People in four tents.

(45)   *Ms. V.:*   Okay, is that right? People in a tent, times four, what times ... four tents times people ... what? [*Writing the labels on the picture*]

(46)   *Linda:*   Four tents times one tent.

(47)   *Ms. V.:*   What's that mean?

(48)   *Ethel:*   I was just thinking, if you were still dividing that way, that would be 12 scouts by four tents.

(49)   *Ms. V.:*   Okay, twelve scouts divided by four tents. It gives you three ...

(50)   *Ethel:*   Three scouts per tent.

(51)   *Ms. V.:*   Three scouts per tent. So, now do this one. Four tents times ...

(52)   *Ethel:*   Three scouts per tent.

(53)   *Ms. V.:*   Three scouts per tent equals ...

(54)   *Ethel:*   Twelve scouts.

The use of labels for each of the numbers involved in the problem helped students recognize the relationships involved and the different roles the ratio and the scale factor have in a proportional situation. The number of scouts and the number of tents are the measure spaces involved. The ratio corresponds with the number of scouts per tent, 3, and the scale factor with the number of groups, 10. By using the units the number sentences take a whole new meaning. The solution on the board is like this.

12 scouts ÷ 4 tents = 3 scouts per tent   4 tents × 3 scouts per tent = 12 scouts

$$40 \times 3 = 120 \text{ scouts}$$
$$4 \text{ tents} \times 10 = 40$$
$$12 \times 10 = 120$$

Of course, the discussion did not finish here. Lee, a very active and participatory listener, had an insight into how those two solutions are related to each other.

(55)   *Ms. V.:*   Lee?

(56)   *Lee:*   Well, I just saw something. You could make this one, or that one into like, um, like a box. Between the 40 and the 4 you could put times 10, and between the 120 and the 12 you could put times 10.

(57)   *Ms. V.:*   Okay, say that again. You noticed something.

(58)   *Lee:*   I noticed something. If you put … between the 4 and the 40, you put times 10.

(59)   *Ms. V.:*   I'll put a little times 10 in there.

(60)   *Lee:*   And then you can put that between the 120 and the 12. And then on the other one you put 4 times 3 is 12, and then you put 40 times 3 is 120.

4   tents × 10 groups (of 4) = 40 tents
$$\times \qquad\qquad \times$$
3 scouts per tent   3 scouts per tent
$$= \qquad\qquad =$$
12 scouts × 10 groups (of 12) = 120 scouts

(61)   *Ms. V.:*   Wow, what do you think about that?

(62)   *Lee:*   A ratio box!

(63)   *Ms. V.:*   A ratio box! What a neat idea. It all fits. Does it all fit together? Rory, you look puzzled. Do you have a question?

(64)   *Rory:*   No, I don't. I see what Lee is saying. You use the 10 to get across, and then the four times the 3 to get down, and then use the opposite on the other one.

(65)   *Ms. V.:*   What does this 10 mean? [*Pause*] Anybody? What do you think, Julia?

(66)   *Julia:*   Maybe like, ... maybe could've see how many groups you could put into? ... Like, there is 10 groups in 120 for 12 and for 40 there is 10 groups of 4.

The way Lee projected his meaning of ratio into the situation was very rich. He not only put together the pieces from the two solutions devised previously but he also connected them with the overall goal of understanding ratios. In earlier classes he had been wondering about what Ms. V. meant by the ratio of a problem, and now he finds his answer. Wenger (1998) has noted the special moment in which a particular understanding leads to deeper meaning, the process of reification. "A certain understanding is given form. This form then becomes a focus for the negotiation of meaning" (p. 59). By visualizing the multiplicative relations involved in a direct proportion as a box, Lee and his classmates now had a tool to help them think about ratios. The process of negotiation continued.

(67)   *Karen:*   I was going to comment what a ratio is. I read in the dictionary, and it said when you compare two things.

(68)   *Ms. V.:*   When you compare two things.

(69)   *Tim:*   You could like label it as scouts and tents and draw lines.

(70)   *Ms. V.:*   Draw lines like this? But did we have to build up on a ratio table to get our answer? What did we use to get our answer? Did we build up four, twelve, and then maybe eight and 24? What did we use to get to our answer?

Tents	4	× 10	40
Scouts	12	× 10	120

(71)   *Karen:*   We multiplied by ten.

(72)   *Ms. V.:*   We multiplied by ten. We used multiplication. Does that make sense? Do you think that you could use a multiplication strategy? Which is more efficient? The ratio tables, or the pictures, or the multiplication strategies?

(73)   *Lee:*   Or this?

(74)   *Ms. V.:*   Or what's the other one, ratio box? Which is most efficient of the ways we've tried and the things you've seen? Julia?

(75)   *Julia:*   I think it might be what we did for this problem.

(76)   *Ms. V.:*   For this problem?

(77)   *Julia:*   Well for a ratio table you usually do, like, four tents for twelve scouts, then you go, eight scouts for 24, and then like double it instead of like timesing it. And this is like really fast.

The last part of the discussion showed again how students made connections among different strategies. Ratio tables and pictures were compared with other, more-efficient strategies helping less-advanced students recognize their own solution in the discussion. In this way, participation is not restricted to actions—it involves connections. Being able to see how your own understanding reflects on others experiences is also a form of participation (Wenger 1998).

After the discussion Ms. V. gave two new problems to work individually. She asked the students to work in pairs and choose the strategy they considered most efficient to present to the whole group. The numbers in these problems were chosen to drive the use of more efficient strategies. Her goal was not to get all students to solve problems using the multiplicative strategies but to help them understand how these strategies worked, how different strategies were connected with their own, and hopefully, to motivate them to try some of the strategies out.

CONCLUDING COMMENTS

This episode in Ms. V.'s classroom shows how the cognitive development of mathematical content is interrelated with the social interactions in the classroom. Lamon (1999) defines proportional reasoning as "the ability to recognize, to explain, to think about, to make conjectures about, to graph, to transform, to compare, to make judgments about, to represent, or to symbolize" direct and inverse proportions (p. 8). In this classroom we have seen how a group of students used a direct-proportion situation to reflect and advance their understanding of ratios and proportion. The social and the sociomathematical norms (Cobb 1995) of the classroom were essential to allow this process. The students' obligation to explain their reasoning and to listen and to make sense of others' solutions characterized this classroom. The teacher played an active and central role. Ms. V. provided direction and guided the development of the class by selecting appropriate tasks, providing relevant information, helping students articulate their methods, and supporting students' thinking.

REFERENCES

Carpenter, Thomas, Elizabeth Fennema, and Megan Franke. "Cognitively Guided Instruction: a Knowledge Base for Reform in Primary Mathematics Instruction." *Elementary School Journal* 97, no. 1 (1997): 3–20.

Cobb, Paul. "Mathematical Learning and Small-Group Interaction: Four Case Studies." In *The Emergence of Mathematical Meaning: Interaction in Classroom Cultures*, edited by Paul Cobb and Heinrich Bauersfeld, pp. 25–129. Hillsdale, N.J.: Lawrence Erlbaum Associates, 1995.

Greeno, James, Allan Collins, and Lauren Resnick. "Cognition and Learning." In *Handbook of Educational Psychology*, edited by David C. Berliner and Robert C. Calfee, pp. 15–46. New York: Macmillan, 1996.

Lamon, Susan. *Teaching Fractions and Ratios for Understanding: Essential Content Knowledge and Instructional Strategies for Teachers*. Mahwah, N.J.: Lawrence Erlbaum Associates, 1999.

O'Connor, Mary Catherine, and Sarah Michaels. "Shifting Participant Frameworks: Orchestrating Thinking Practices in Group Discussion." In *Discourse, Learning and Schooling*, edited by Deborah Hicks, pp. 63–103. New York: Cambridge University Press, 1996.

Wenger, Etienne. *Communities of Practice: Learning, Meaning, and Identity*. New York: Cambridge University Press, 1998.

24

Fraction Instruction That Fosters Multiplicative Reasoning

Lee S. Vanhille

Arthur J. Baroody

FRACTIONS and proportions, although difficult for many students, are important, complex mathematical ideas. Operations on fractions and proportional reasoning have traditionally been treated as two separate topics. Because understanding both topics involves multiplicative reasoning, fraction instruction that fosters such reasoning should also promote proportional reasoning. In this article, we shall describe the rationale for such a program of instruction, how it was implemented, and the program's impact on sixth graders' multiplicative reasoning, including their proportional reasoning.

RATIONALE FOR THE INSTRUCTIONAL PROGRAM

Why does a deep understanding of fractions and proportions require multiplicative reasoning, and why does traditional instruction often fail to promote such reasoning?

The Crucial Role of Multiplicative Reasoning

Fractions. Fractions involve both a *between*-fractions multiplicative relation and within-fraction multiplicative relations. The former defines the relation between the numerators and between the denominators of equivalent fractions. For 3/5 and 6/10, the between-fractions relation is a factor of 2 because the numerator of the latter, 6, is twice that of the former, 3, and the denominator of the second, 10, is twice that of the first, 5—that is,
$$3/5 = (3 \times 2)/(5 \times 2) = 6/10.$$

In traditional elementary school programs, between-fractions multiplicative factors are typically integers (as the example above illustrates) and only rarely nonintegers, such as

$$4/6 = (4 \times 1\ 1/2)/(6 \times 1\ 1/2) = 6/9 \ .$$

The *within*-fraction relations define the connections between a fraction's numerator and denominator. For the fraction 3/5, the numerator, 3 is 3/5 of the denominator, 5 (i.e., $3 = 3/5 \times 5$); and the denominator, 5, is 5/3 of 3 (i.e., $5 = 5/3 \times 3$). Furthermore, all equivalent fractions share the same within-fraction relations. For instance, as with 3/5, the numerator 6 of the equivalent fraction 6/10 is 3/5 of the denominator 10 ($6 = 3/5 \times 10$). Conversely, if two fractions have the same within-fraction relations, they must be equivalent. Except to simplify improper fractions (e.g., $5/3 = 1\ 2/3$) by a memorized procedure, within-fraction multiplicative relations are typically not discussed in elementary school.

Proportions. Working with proportions is in many ways similar to working with equivalent fractions. Consider the case of similar rectangles. If two similar rectangles have the dimensions 10×6 and $? \times 9$, then the width of the second is 1 1/2 times as large as the width of the first ($6 \times 1\ 1/2 = 9$). Therefore, the length of the second rectangle is 1 1/2 times as large as the length of the first ($10 \times 1\ 1/2 = 15$). Alternatively, the unknown length could be determined by comparing width to length. Since the length of the first rectangle is 1 2/3 as large as its width ($6 \times 1\ 2/3 = 10$), it follows that the length of the second rectangle is also 1 2/3 as large as its width ($9 \times 1\ 2/3 = 15$).

Sources of Learning Difficulties

Fractions. Many children have difficulty understanding fractions, including operations on fractions. One possible reason for this is that students lack the concrete experiences necessary to construct conceptual understanding, or they do not see how abstract procedures are tied to concrete experiences. A second possible reason is that students are relatively unfamiliar with the multiplicative reasoning that is required to understand fractions, a gap that instruction does not adequately address. (See for example, Baroody with Coslick [1998], for a discussion of common difficulties.)

Proportions. Many students have difficulty with proportions because traditional instruction on this topic does not promote an understanding of multiplicative relations. Instead, students are taught to solve proportion problems by applying a rotely memorized cross-multiplication algorithm. Because many students either never learn or forget this meaningless procedure, they may attempt to solve proportion problems by using additive reasoning (e.g., responding to the proportion $2/3 = x/9$ by reasoning that if 6 must be added to 3 to make 9, then 6 added to 2 is 8, and so *x* is 8). Typically, even successfully memorizing and using the cross-multiplication procedure does not challenge students' additive thinking or promote multiplicative reasoning.

Moreover, traditional instruction does not capitalize on the structurally similar nature of fractions and proportions. It seldom helps students recognize, for example, that their procedures for working with fractional equivalents are relevant to solving proportion problems (Heller et al. 1990). As Heller and her colleagues (1990) noted, "Rational number skills, once developed, are not being used or applied to areas of obvious application [such as proportions]. Because the rational number skills appear first in the school curriculum, they could be used as the computational framework within which proportional relationships could be investigated" (p. 400).

A Program of Fraction Instruction That Promotes Multiplicative Reasoning

We describe here a program of fraction instruction to address the deficiencies of traditional fraction and proportion instruction just discussed. More specifically, this program uses the familiar analogy of fair sharing and concrete or pictorial models to help students conceptually understand operations on fractions, including the multiplicative reasoning required for such understanding. By promoting more sophisticated multiplicative reasoning within the context of fractions, students should improve their understanding of proportion problems and their ability to devise strategies spontaneously for solving them.

Distinctive Characteristics of the Program

Six features distinguish this instruction from traditional approaches.

1. Fractions and operations on fractions are linked in the program to a seldom-discussed *operator* meaning of rational numbers, that is, taking a fractional part of some amount. With this meaning, a fraction *a/b* can be interpreted as *a* out of *b* equal groups that total some amount of items. For example, 2/3 of 12 cookies is thought of as two of three equal groups that total twelve cookies.

2. A concrete model that can be treated as either a collection (a discrete quantity) or an amount of area (a continuous quantity) is used to help make this meaning clear by relating it to something real. Specifically, equal-sized pieces of paper are used to represent various objects, such as cookies. Using

"cookies" as a model provides instructional flexibility in two ways. One is that "cookies" can be treated as a discrete quantity and can easily be counted. The other is that individual cookies can also be treated, if need be, as a continuous (area) quantity and subdivided (partitioned).

3. The operator interpretation of fractions builds on children's informal understanding of fair sharing, and modeling this analogy with "cookies" involves *equal partitioning*, a conceptual basis for fractions. For example, 2/3 of 12 cookies was defined as sharing twelve cookies among three people and then determining how many cookies two of the people had. In effect, the denominator indicates the number of groups, 3, into which the 12 cookies are distributed, resulting in 4 cookies for each group. The numerator indicates that the total number of cookies from 2 groups should be counted; hence there are 8 cookies in the 2 groups. Therefore, 2/3 of 12 is 8. The equal partitioning increases students' proficiency with operations on fractions and builds a foundation for more difficult multiplicative skills.

4. Equal-partitioning experiences build the foundation for dealing with *noninteger factors* and *complex fractions*, which are not part of typical fraction instruction. Students, for example, dealt with 3/4 of 10 (also represented as 3/4 = ?/10) by distributing 10 cookies into 4 groups, resulting in 2 1/2 cookies in each group (the noninteger factor in the between-fractions relation) and 7 1/2 cookies in 3 groups; that is, 3/4 of 10 is 7 1/2, which can also be represented 3/4 = (7 1/2)/10. Note that the second ratio in this equation is a complex fraction.

5. To provide further experience with noninteger factors, instruction in this program includes experience with an informal method for multiplying a mixed number by a whole number—a method based on a *distributive model*. The emphasis during instruction is on mental computation. Consider, for example, the following task: (2 1/3) × 3. This task might be expressed as, "Two threes plus one-third of three. Two threes are six, and one-third of three is one; therefore, 2 1/3 × 3 = 6 + 1 = 7."

Next, this skill is applied to the within-fraction and between-fractions relations in nontraditional equivalent fractions, as in the task 6/8 = 9/?. In the first fraction, the denominator is one-third larger than the numerator, so the same within-fraction relation must be true in the second fraction: one-third of 9 is 3; 9 + 3 = 12; therefore, 6/8 = 9/12. The between-fractions relation is also a noninteger factor. The numerator of the second fraction is 1 1/2 times as large as the numerator of the first, so the same relation must be true between the denominators. Therefore, 1 1/2 × 8 is 8 plus half of 8; 8 + 4 = 12, and so 6/8 = 9/12. Students can determine the factors either by recognizing these number relations (after practice with multiplication using the distributive model) or by partitioning. With partitioning, in the previous problem the student interprets 6/8 as 6 of 8 groups; in those 6 groups there are 9 objects, which results in 1 1/2 objects in each group (the between-fractions relation). Since all 8 groups contain 1 1/2 objects per group, there is a total of 12 objects.

6. *Fraction tables are used to examine the relations among families of fractions* generated by partitioning experiences. For example, 3/4 of 12 is 9, which can also be written 3/4 = 9/12. Taking 3/4 of other amounts, such as 3/4 of 8, 12, 16, and 20, students generate the fraction table 3/4 = 6/8 = 9/12 = 12/16 = 15/20. Students then use their table to explore within-fraction relations (the denominator is always one-third larger than the numerator, and the numerator is always three-fourths of the denominator) and between-fractions relations (e.g., as the numerator increases by 3, the denominator increases by 4).

Implementation of the Program

Fair-sharing experience and notation. During the first day of the instruction, students are introduced to an "operator" meaning of fractions. This initial instruction is based on equal-partitioning experiences in the form of solving "fair sharing" problems with a concrete model (e.g., "cookies"). After three or four days of fair-sharing experiences, students justify all subsequent work, including quizzes and test, with drawings to represent the partitioning. For example, in one class 3/4 of 8 ("Take eight objects, evenly distribute them into four groups, and count the number of objects in three groups") was represented on paper as

The student first drew four lines to represent the four groups and then wrote in a 2 above each line to represent the result of divvying-up eight objects among the four groups. Three of the four 2s were then circled to represent three of the four groups. Students used this notation in all their fraction work, including equivalency, addition, subtraction, multiplication, and division.

Fraction equivalents. These fair-sharing experiences are linked to fraction equivalents. For example, in the written representation for 3/4 of 8 is 6, namely,

the relation can be summarized symbolically by the equation 3/4=6/8. By using this procedure, students can construct families of equivalent fractions. For instance, they can find 2/3 of different quantities and display their results in the form of a fraction table, as previously discussed. For example, 2/3 of 3, 6, 9, and 15, in turn, would result in the equivalent fractions

$$2/3 = 4/6 = 6/9 = 10/15.$$

Classroom discussion can center on patterns and relations. The students might notice that the numerators increase by 2 and the denominators increase by 3. Given this pattern, they may notice a "hole" where the fraction 8/12 should be. They can confirm this conjecture by using a drawing to find 2/3 of 12 (12 cookies distributed fairly among 3 groups yields 4 cookies in each group; 2 of the 3 groups contain 8 of the 12 cookies). A teacher might also ask students what else these fractions have in common. One speculation might be that the numerators are all 2/3 of the denominators, which can be confirmed with drawings. A more difficult question is this: "If the numerators are 2/3 of the denominators, what are the denominators in terms of the numerators?" In this instance, the denominators are "half again as big" as (or one-half larger than or 1 1/2 times as large as) the numerators. A teacher can add another step in the notation by indicating the number of cookies in each group; this number is the multiplicative factor between the equivalent fractions.

Early in the process, students can deal with quantities of cookies that are not multiples of the denominators, such as 2/3 of 10. When a teacher insists that the three people receive the same amount of cookies—and that all the cookies must be shared—some students may break the leftover cookie into thirds. A student's solution may look like the following:

There are 3 1/3 cookies in each group, and in 2 groups there are 6 2/3; therefore, 2/3 = (6 2/3)/10, which is a complex fraction with a noninteger multiplicative factor. This equivalency may be represented as:

$$(2 \times 2\ 1/3) / (3 \times 2\ 1/3) = (6\ 2/3) / 10$$

As a result of their instruction, students should have many strategies for determining whether 6/8 equals 9/12. A student could interpret the relation as 6/8 of 12, for example, requiring 12 items to be shared among 8 groups, or 1 1/2 items per group. If 6 groups each have 1 1/2 items, then there would be 9 items ($6 \times 1 = 9$). The student might also recognize, without partitioning, the between-fractions factor of 1 1/2 or the within-fraction factor of 1 1/3 ($6 \times 1\ 1/3 = 8$). Alternatively, the student might discover that both fractions reduce to 3/4, which is a second within-fraction factor. That is, the numerators are 3/4 of the denominators. This can be justified through partitioning activities to show that 3/4 of 8 is 6 and that 3/4 of 12 is 9.

Multiplication. One approach to the multiplication of fractions that fits well into this fraction program is to represent multiplication as successive partitioning actions. The purpose of this particular representation is to provide further practice with partitioning and to explore multiplicative relations. For example, 3/4 × 2/3 is defined as "3/4 of 2/3 of some amount." If we start with 12 cookies, 2/3 of 12 is 8. We then do 3/4 of 8, which is 6. Thus, 3/4 × 2/3 = 6/12 = 1/2. Six-twelfths is the ratio of the final amount, 6, compared to the starting amount, 12. (Note that the fractions had different referent units. The referent unit for 2/3 is 12 cookies, whereas the referent unit for 3/4 is 8 cookies.) An informal representation may be this:

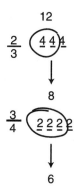

After these first experiences, students can deal with multiplication problems in which they are not given the initial number of cookies. The task becomes more challenging when they must find the smallest number of cookies that will work without fractional parts. Even more challenging is finding the smallest number of cookies that result in whole cookies for the outcome but that allow for fractional parts during the partitioning process. With the problem 3/4 × 2/3, a class might construct the following fraction table from their collaborative investigations:

1/2 = 2/4 = 3/6 = (4 1/2)/9 = 6/12 = 9/18 = 10/20 = 12/24.

Students should readily see the relatively easy within-fraction relation: The numerators are one-half of the denominators, and the denominators are twice the numerators.

Other multiplication problems can allow further opportunity to investigate more complicated multiplicative relations. For 2/3 × 3/5, a fraction table might be 2/5 = 4/10 = 6/15 = 8/20 = 12/30. Students can then be asked, "What are the within-fraction relations in the fraction table for 2/3 × 3/5?"

With class discussion, students can realize that the denominators are two and a half times as large as the numerators and the numerators are 2/5 of the denominators.

In a different aspect of multiplication, the students deal with the multiplication of a mixed number by a whole number using a distributive model (as explained previously) instead of the traditional approach of changing mixed numbers to improper fractions and multiplying according to the procedure for multiplying simple fractions. This experience can help students recognize multiplicative relations between numbers in equivalent fractions and fraction tables.

In summary, the instructional program is designed to teach fraction computation as well as enhance students' general multiplicative reasoning. Principal features include equal-partitioning (fair sharing) experiences and drawings to justify all work. Other important features are experience with noninteger multiplicative factors and complex fractions, fraction tables, and a distributive model for the multiplication of a mixed number by a whole number. Thus, increasingly difficult multiplicative relations are explored within the framework of fractions.

THE IMPACT OF THE PROGRAM

The experimental instructional program described above was evaluated with one class of sixth graders. Two classes that received traditional textbook instruction served as comparison groups. Four measures were used as a pretest and posttest for students in all three classes: (1) fraction computation (equivalent fractions with integer factors and the addition, subtraction, multiplication, and division of simple fractions); (2) equivalent fractions with noninteger factors (9/12 = 12/?); (3) missing-factor sentences with noninteger factors (e.g., 4 × ? = 10); and (4) proportion problems. Below, we briefly discuss the results regarding the first three tasks and then focus on the results of the last task, which gauged students' transfer (application) of multiplicative reasoning to a new topic. We end the article by discussing the educational implication of the program's evaluation.

Results Regarding Fractions and Operations on Fractions

The experimental instructional program was as effective as a traditional one in fostering the mastery of basic computational skills. Moreover, the experimental instruction was significantly more successful than the traditional approach in promoting proficiency with equivalent fractions with noninteger factors and in solving missing-factor sentences with noninteger factors. The missing-factor sentences involved the application of multiplicative reasoning to an untaught situation.

Results Regarding Proportion Problems

All students were administered five tasks in each of four different contexts, for a total of twenty proportion problems. In the Eels-and-Goldfish context, the participants were given problems such as this: "If a 6 cm eel eats 12 goldfish, how many goldfish do you think a 4 cm eel would eat?" (In subsequent examples of proportionality problems in the remainder of this article, we will sometimes use the notation $a:b::c:d$. Thus, this example could be represented 6:12::4:?. This notation was not used with the student but is used here to help the reader distinguish between the instructional fraction tasks and the proportionality problems of the interviews.) In the Food-Boxes context, the students were given problems like this one: "If on a camping trip 12 people eat 16 boxes of food, how many boxes would 15 people eat?" In the Similar-Rectangles context, the participants saw a pair of similar rectangles, the length and width of one rectangle, and only one dimension of the other. They were instructed to find the length of the missing side. In the Number-Rule context, the students were shown two pairs of numbers, such as 6 ⟶8 and 12 ⟶16, and asked to explain rules that could change the first number into the second number. They were then given the opportunity to apply their rules to subsequent number pairs, such as 15 ⟶?.

How students receiving the experimental instruction compared to those receiving traditional instruction. Performance on the proportion problems was evaluated according to a six-point proportionality scale. The top two levels indicated multiplicative reasoning; the middle two levels indicated a transitional phase (such as combining multiplicative and additive reasoning, as in doubling and adding 5 in the problem 6:15::10:?, in which the student determines the unknown to be 25); the lowest two levels indicated incorrect reasoning, such as illogical or additive reasoning. Students who received the experimental instruction used strategies based on multiplicative reasoning significantly more often than students who received traditional instruction.

Examples of multiplicative reasoning by students who received the experimental instruction. The first example is illustrative of those participants who moved from using relatively low-level reasoning on the proportion pretest to using transitional reasoning on the posttest. On the pretest, Amber typically used additive reasoning and was correct on only three of twenty proportion problems. For the Similar-Rectangles problem involving an 8×12 rectangle, for example, she reasoned that the length was 4 longer than the width and, given that the width of the second rectangle was 12, its length must be 12 + 4, or 16. Amber was successful, though, with integer factors in three of the four contexts. In the Eel-and-Goldfish context, for instance, she determined that if 6 eels eat 12 goldfish, 4 would eat 8. For the next problem (If 10 eels eat 25 goldfish, then 6 will eat how many?), which involved a noninteger factor, Amber made several unsuccessful attempts to use multiplicative reasoning. Specifically, after noting that 10 times 2 was 20, she concluded, "But it eats 25 goldfish. It needs 5 more, and 10 × 3 was too much." At this point, she was stumped and made no further attempt to solve the problem.

On the posttest, Amber typically used multiplicative reasoning in two of the four contexts, including problems that involved noninteger factors; she successfully solved eight of twenty proportion problems. For the Food-Boxes problem that involved the proportion 8 is to 20 as 10 is to x, for example, she reasoned that each person required 2 1/2 boxes and so 10 people would require 25 boxes. The interview excerpt below reveals an apparently new ability to partition:

Interviewer: Tell me first of all, how did you get the 2 1/2 each?

Amber: I put [i.e., divided] 20 into [i.e., among] 8 people.

Interviewer: Okay, and how do you do that?

Amber: They each get 2, and there's 4 left over, then [I] cut each in half.

Each person received half of a box (in addition to 2 whole boxes) through her partitioning activities, but Amber may not have realized the multiplicative relation that the leftover 4 was half of 8. In the same context, she also successfully used partitioning on three other problems, including one involving 5:6::6:x. Amber also got the majority of the Eels-and-Goldfish problems correct because this context, like the Food-and-Boxes context, entailed representing concrete situations involving collections or discrete quantities. (Her lack of success on the Number-Rule context and the Similar-Rectangles context may have been due to the fact that the former simply involved numbers and was relatively abstract and that the latter involved measures of length or continuous quantities.)

The second example involves a participant who was solidly in the transitional phase when the study began and advanced to multiplicative reasoning by the end of the study. On the pretest, Janika reasoned multiplicatively with integer factors in all four contexts and solved eight of twenty problems. Moreover, in the Food-Boxes context, she was even able to use partitioning skills (which few students exhibited on the pretest). Figure 24.1 illustrates her solution to a problem involving the proportion 12:16::15:x. She circled 12 of the 16 boxes, which represented one box for each person. With the four leftover boxes, she apparently mentally calculated one-third box for each person. She represented 15 people, who would require one box each; she then drew lines through every three people, who shared one box among them, for a total of 1 1/3 boxes each.

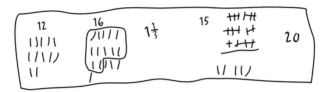

Fig. 24.1. Partitioning used to solve 12:16::15:x

On the posttest, Janika showed competent multiplicative reasoning on all twenty proportion problems. Using partitioning to determine the first relation and the distributive model to confirm it, she then applied the distributive model to generate other number pairs. For example, in the Similar-Rectangles problem 12:15::16:?, Janika made 12 groups, distributed one in each, divided the extra three into fourths, and distributed one-fourth into each group (1 1/4 in each of the 12 groups). Using the distributive model, she multiplied (1 1/4) × 12 to confirm the partitioning and then multiplied (1 1/4) × 16 to determine the answer, 20 (see fig. 24.2).

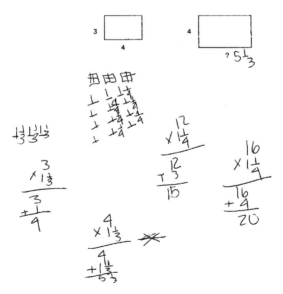

Fig. 24.2. Strategies to solve 12:15::16:

As a result of the experimental instruction, Janika apparently expanded her partitioning abilities and acquired the ability to use the distributive model for the multiplication of mixed numerals.

EDUCATIONAL IMPORTANCE

Instruction on fractions, including operations on fractions, does not have to be confusing to students. It can be made meaningful by using an understandable analogy and by building on children's existing informal and formal knowledge (e.g., Mack [1993]). Although there are several ways this can be accomplished (see, for example, Baroody with Coslick [1998]), the experimental instruction described in this article has a number of benefits:

1. It can serve to introduce students to an often overlooked meaning of rational numbers, namely, the operator (a fraction of an amount) meaning. In order to develop a deep understanding of fraction symbolism and rational numbers, students need to recognize that fractions can represent more than a part-whole meaning.

2. Students can make sense of the instruction because it relates the operator meaning to the informal and meaningful analogy of fairly sharing (equally partitioning) a collection, situations they easily model with discrete but subdivided quantities such as "cookies." For example, 3/4 of 8 can be thought of as 8 items divided into 4 groups and taking 3 of the 4 groups.

3. Particularly important, the instruction can help students bridge the gap between additive and multiplicative reasoning in two ways. Children gain experience with noninteger factors through equal-partitioning activities and using a distributive model for multiplying a mixed number by a whole number. The instruction also fosters multiplicative reasoning by focusing students' attention on both a between-fractions relation, which underlies equivalent fractions (e.g., 3/4 = (3 × 2) / (4 ×2) = 6/8), and within-fraction relations, the functions that describe the relations between the numerator and denominator of a fraction and all equivalent fractions (e.g., for 3/4, 9/12, and so on, the numerator in each fraction is 3/4 of the denominator).

Because of the last two points above, the experimental fraction instruction discussed in this article can help children master basic fraction skills, such as determining equivalent fractions and operating on fractions, in a meaningful fashion. Furthermore, because this instruction promotes multiplicative reasoning, students can successfully tackle other novel tasks that involve such reasoning, including proportion problems. Thus, it can provide a basis for meaningfully or adaptively engaging in numerous mathematical, scientific, or everyday tasks such as dealing with similar rectangles; interpreting the slope of a line; and working with percentage and variables like speed, acceleration, density, concentration, or intensity.

REFERENCES

Baroody, Arthur J., with Ronald T. Coslick. *Fostering Children's Mathematical Power: An Investigative Approach to K–8 Mathematics Instruction*. Mahwah, N.J.: Lawrence Erlbaum Associates, 1998.

Heller, Patricia M., Thomas R. Post, Merlyn Behr, and Richard Lesh. "Qualitative and Numerical Reasoning about Fractions and Rates by Seventh- and Eighth-Grade Students." *Journal for Research in Mathematics Education* 21 (November 1990), pp. 388–402.

Mack, N. K. "Learning Rational Numbers with Understanding: The Case of Informal Knowledge." In *Rational Numbers: An Integration of Research* , edited by Thomas P. Carpenter, Elizabeth Fennema, and Thomas A. Romberg, pp. 85–105. Hillsdale, N.J.: Lawrence Erlbaum Associates, 1993.

25

Profound Understanding of Division of Fractions

Alfinio Flores

A PURPOSE of this article is to illustrate some of the mathematical knowledge that teachers need so that the teaching of division of fractions is meaningful for the students. Ma (1999) emphasized four crucial properties of understanding—connectedness, multiple perspectives, basic ideas, and longitudinal coherence. Few comprehensive discussions of division of fractions are readily available for today's teachers. There are presentations of division of fractions in terms that are meaningful for students in the upper elementary and middle school, using the measurement interpretation of division, or dividing fractions with equal denominators (Armstrong and Bezuk 1995). This is a good start, but it is not enough. Teachers need a complete picture that connects concrete approaches of division with the algorithm of multiplying by the reciprocal. They need to understand the role of reciprocals (multiplicative inverses) and the inverse nature of the operations of division and multiplication.

Traditionally, in the United States, division of fractions has been taught often by emphasizing the algorithmic procedure "invert the second fraction and multiply" with little effort to provide students with an understanding of why it works. Sometimes teachers and students do not realize that the "invert" refers to the multiplicative inverse, that is, the reciprocal; rather, they see the instruction to invert only as a symbol manipulation where the fraction is "flipped."

Many of the concepts needed to understand division of fractions are based on relations among whole numbers that are multiplicative in nature. These concepts need to be developed before the topic is treated. Understanding these relations is important not only for the teachers in the upper elementary and middle school but also for teachers of earlier grades.

Connectedness

Teachers who understand a topic make connections with other mathematical concepts and procedures. They also emphasize more-complicated and underlying connections among different mathematical operations and subdomains (Ma 1999). Some of the connections needed in division of fractions are fractions and quotients, fractions and ratios, division as multiplicative comparison, reciprocals (inverse elements), and inverse operations. These connections will help students' learning be coherent. Instead of learning division of fractions as an isolated topic, students can learn how it fits in a unified body of knowledge. Division of fractions also provides a setting to develop proportional thinking, which is at the core of mathematics in the middle school. Students also have the opportunity to develop their algebraic thinking by dealing with concepts such as inverse operation and reciprocal.

Multiple Perspectives

Meanings of Division of Fractions

Understanding division of fractions is helped by to appreciating different meanings such as measurement division, sharing, finding a whole given a part, and missing factors. Instruction can be improved if we value various approaches to a solution and different kinds of explanations and are aware of advantages and disadvantages and the contexts in which each approach tends to be more helpful.

One meaning of division is the measurement interpretation: How many times does one number fit into another? With the help of concrete models of fractions, students can see that $1/4$ fits two times into $1/2$, therefore $1/2 \div 1/4 = 2$. With some guidance, they will also be able to solve problems like $1/6 \div 1/3 = 1/2$, and $1/2 \div 1/3 = 1\ 1/2 = 3/2$ (fig. 25.1). Sewing or cooking are two contexts commonly used in the measurement interpretation of division as in this example: "You have 1 3/4 cups of flour, and for each batch of cupcakes you need 1/2 a cup. How many batches can you make?"

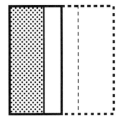

Fig. 25.1. 1/3 fits one and a half times into 1/2.

Another meaning is sharing or partitive division. Division as sharing works well when fractions are divided by whole numbers. To solve 1/2 ÷ 3 = 1/6, one half is divided into three equal parts, and the result is 1/6 of the original whole, for example, when half a cake is shared equally among three children.

One can also interpret division of fractions as a problem of finding a whole given a part. A teacher explained "1 3/4 ÷ 1/2 means that 1/2 of a number is 1 3/4. The answer as one can imagine, will be 3 1/2 which is exactly the same as the answer for 1 3/4 × 2. 2 is the reciprocal of 1/2" (Ma 1999, p. 60).

A problem like 1 3/4 ÷ 1/2 can also be interpreted as a missing factor problem: "What number multiplied by 1/2 gives 1 3/4?" Although the previous interpretation is similar to this one, they are not exactly the same. The missing factor meaning of division is also used with whole numbers, 2 times what gives 6? However "finding a whole given a part" does not work for division problems with whole numbers.

We can think of fractions of the same denominator as sets made of pieces of the same size. Students also need to understand the relation between quotients and fractions, that is, 2 ÷ 3 = 2/3 . We can use the meaning of division as a multiplicative comparison between the number of pieces in two sets, that is, as a ratio. When the pieces are the same size, the ratio depends only on the number of pieces involved, not on their size. Thus the problems 2 ÷ 3, 2/5 ÷ 3/5, and 2/8 ÷ 3/8 all have the same answer, 2/3. To divide fractions with same denominators, it is enough to divide the numerators.

Students' Methods

Teachers can learn from students' invented methods, adapt these methods, and help students see why their methods work. I will present three examples of methods used by students. In the first example, a student (in a class of ages 11–12) used an approach that involves inverse proportionality (as well as measurement interpretation). To solve 1/2 ÷ 2, his thinking was "there is only half a two in one, so there is a quarter in half of one" (Pirie 1988, p. 3).

One teacher expanded a method used by students: "When working on whole numbers my students learned to solve certain kind of problems in a simpler way by applying the distributive law. This approach applies to operations with fractions too" (Ma 1999, p. 63). To calculate 1 3/4 ÷ 1/2, write 1 3/4 as 1 + 3/4, divide each part by 1/2, and add the two quotients:

$$(1 + 3/4) \div 1/2 = (1 \div 1/2) + (3/4 \div 1/2) = 2 + 1\ 1/2 = 3\ 1/2.$$

A ten-year-old girl developed a procedure to divide by fractions whose numerator is one less than its denominator using a part-complement representation (Rowland 1997). The procedure can be described symbolically as

$$100 \div 3/4 = 100 + 100 \div 3$$
$$100 \div 4/5 = 100 + 100 \div 4.$$

I will use figure 25.2 to explain this procedure for 100 ÷ 3/4. The process has two steps. First, we see that 3/4 goes into 100 units, 100 times (the shaded areas). For each circle, there is a remainder of 1/4. Then we need to divide 3 pieces of 1/4 into the remaining 100 pieces of 1/4. That gives us the additional 100 ÷ 3.

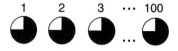

Figure 25.2. 100 ÷ 3/4

As these examples illustrate, students' thinking can be quite original and not always easy to understand. Profound understanding of fractions will allow teachers to make sense of these procedures and help students make connections to other procedures and concepts.

Ways of Justification

Teachers can use concrete representations, empirical evidence and patterns, and properties of numbers and operations to explain the various approaches to division of fractions. Furthermore, by using the basic ideas and properties of fractional numbers and operations, teachers will help their students develop analytical schemes of proof.

The teacher can present multiplication problems and division problems that are easy to solve using concrete or pictorial representations, and propitiate the recognition of patterns when students record the results in a systematic way. Students should notice the reciprocals in each row, such as 1/3 and

3. They can then explore whether the pattern holds for other problems, such as $1/2 \div 1/3$.

$2 \div 1/3 = 6$	$2 \times 3 = 6$
$2 \div 1/4 = 8$	$2 \times 4 = 8$
$2 \div 1/5 = 10$	$2 \times 5 = 10$

Students can use the particular case of division of fractions when the fractions have the same denominator to develop the procedure for the general case. Given a division problem with fractions such as $3/4 \div 2/5$ they can find equivalent fractions with the same denominator, and then divide the numerators. Students who know how to multiply fractions will then be able to make the connection to the "multiply by the reciprocal" procedure:

$$3/4 \div 2/5 = (3 \times 5)/(4 \times 5) \div (2 \times 4)/(5 \times 4) = (3 \times 5)/(2 \times 4) = (3 \times 5)/(4 \times 2) = 3/4 \times 5/2.$$

Another important special case is when the unit, 1, is divided by a fraction. Students can use measurement interpretation of division to find the answer to problems like $1 \div 2/5$. They can see that $2/5$ fits two and a half times into 1. By using also improper fractions, students will be able to notice that the result is the reciprocal of the number they are dividing by, $1 \div 2/5 = 2\ 1/2 = 5/2$. Students can then use proportional thinking to find the answer for the general case. If $1 \div 2/5 = 5/2$ then $2 \div 2/5 = 2 \times 5/2$, because 2 is twice as big as 1, so the result has to be twice as big. In the same way $3/4 \div 2/5 = 3/4 \times 5/2$, because the result has to be 3/4 as big.

BASIC IDEAS

Teachers with profound understanding of division of fractions are well aware of the basic concepts and principles of fractions and division, such as identity element for multiplication, reciprocals (multiplicative inverses), and the inverse nature of the operations of division and multiplication. As they teach, they revisit and reinforce these basic ideas. These ideas also guide their ways of justification.

The multiplicative identity element, 1, plays a central role. In the context of rational numbers, the number 1 has a double meaning. The number 1 is the unit of comparison, and it is the identity element for multiplication. Multiplying a fraction by 1 does not change the value of the fraction. Students need to see that when multiplying by 1, the identity element is often written as 2/2, 3/3, 4/4, and so on.

Reciprocals need to be stressed. Students need to realize that 1/4 is the unit divided into 4 equal parts, and also that the unit is made of four parts of 1/4 each. Students also need to see that 1/4 of 4 is 1. These reciprocal relations hold of course for all fractions. 5/2 is two and a half units, and also two fifths of 5/2 is 1. Fractions whose product is 1, such as 1/5 and 5, need to be connected in the students' minds.

Children go through several stages to develop the idea of fraction in the context of subdividing areas (Piaget, Inhelder, and Szeminska 1960). Teachers need to make sure students have developed a fairly complete understanding of fractions before discussing division of fractions. As Kieren (1992) points out, a full understanding of fractions in this context requires the ability to partition wholes (or units of different kinds); to reconfigure wholes from parts, which is a psychological basis for the concept of inverse; and to subdivide a part and relate the subdivisions to the part as well as to the original whole, which is a geometric basis for composite operations.

Rational and fractional numbers are at the same time quotients and ratios (Kieren 1992). Fractions like 1/4 and 2/8 are equal in an absolute sense (extensive quantity) and in a relative sense (proportional equality). Students are faced with results that are simultaneously an amount or the result of a division, (1/4 of a cake as one's share) and a ratio (2 cakes for 8 persons, or 1 for 4). Teachers need to understand how the concepts of a fraction 3/4, a quotient 3 ÷ 4, and the ratio 3 to 4 are different and related to each other.

LONGITUDINAL COHERENCE

The knowledge of teachers with profound understanding is not limited to what they teach at a certain grade. Rather, they have achieved a fundamental understanding of fractions and their relation to the other areas in the elementary school mathematics curriculum; they are ready to take advantage of an opportunity to review crucial concepts previously learned by students, such as equivalent fractions, meanings of division with whole numbers, and the relation between multiplication and division. Teachers also know what topics students are going to learn later that are related to division of fractions, and lay the proper foundation for them. Multiplicative thinking (such as comparison of quantities in terms of ratios) and proportional thinking are crucial for students' success in algebra.

Previous Knowledge of Division

A thorough understanding of the operations of division and multiplication with whole numbers is basic for understanding division of fractions. Students need to be familiar with several meanings of division, such as sharing and measurement. Whereas with whole numbers usually both interpretations are possible, with fractions, depending on the problem, usually one interpretation is more helpful than the other. Particularly important is that students make the connection between fractions and quotients. Three cakes shared equally by five children will give 3/5 of a cake to each child (3 ÷ 5 = 3/5). An important special case is when the unit is divided (1 ÷ 5 = 1/5). Sometimes students conceptualize division of whole numbers and fractions as separate and distinct. One student expressed, "We are not talking about fractions, we are talking about dividing" (Toluk 1999, p. 182). Division is also a multiplicative comparison of two quantities, that is, a ratio. The divisions 3 ÷ 5, 6 ÷ 10, and 9 ÷ 15 all have the same result because the ratio in each case is 3 to 5.

Equivalent Fractions

Students need a good understanding of equivalent fractions. This includes seeing that the fractions 1/2, 2/4, 3/6, 4/8 are all equivalent, because they cover the same area in a fraction model, and because the relation of the numerator to the denominator is the same: the numerator is half the denominator. They need to realize that one way to obtain an equivalent fraction is multiplying by 1, written, for example, in the form . Students need facility expressing a fraction as a mixed number or as an improper fraction.

Multiplication and Division

In the same way that subtraction is the reverse operation of addition, division is the reverse operation of multiplication. If we divide and multiply by the same number, the result is the number that we start with; that is, multiplication "undoes" division, (8 ÷ 2) × 2 = 8. This reverse relationship is not limited to cases when the quotient is a whole number, (3 ÷ 5) × 5 = 3. Of course, division also "undoes" multiplication, (8 × 4) ÷ 4 = 8.

This reverse relationship between multiplication and division holds also for fractions: 3/4 ÷ 1/4 = 3, and 3 × 1/4 = 3/4, therefore (3/4 ÷ 1/4) × 1/4 = 3/4. So, when a given fraction is divided by another fraction and then the result is multiplied by the second fraction, we obtain the original fraction, (3/4 ÷ 2/5) × 2/5 = 3/4.

From the last equation we can establish a connection to the "invert and multiply" algorithm. We can multiply both sides by 5/2 to have (3/4 ÷ 2/5) × 2/5 × 5/2 = 3/4 × 5/2. The left side is equal to (3/4 ÷ 2/5) × 1 which is equal to 3/4 ÷ 2/5, therefore 3/4 ÷ 2/5 = 3/4 × 5/2. Notice that in this approach the expression 3/4 ÷ 2/5 is treated as a mathematical object, an important step in the transition from arithmetic to algebra.

Students can also see that 3/4 × 5/2 is equal to 3/4 ÷ 2/5 by using the fact that $a ÷ b = c$ if and only if $c × b = a$. So, to verify that 3/4 × 5/2 = 3/4 ÷ 2/5, multiply 3/4 × 5/2 by 2/5 and see that in fact we obtain 3/4. An important special case for fractions is when reciprocals are multiplied. Because a/b × $b/a = 1$, therefore $1 ÷ a/b = b/a$.

The approach of maintaining the value of a quotient underlies the procedure to divide decimals. To compute $0.25 \overline{)1.75}$, both numbers are multiplied by 100; the answer for $25 \overline{)175}$ is the same. Likewise, for a problem like 1 3/4 ÷ 1/2, when the dividend and the divisor are multiplied by the same number, the result will not change. Thus, 1 3/4 ÷ 1/2 = (1 3/4 × 2) ÷ (1/2 × 2) = (1 3/4 × 2) ÷ 1 = 1 3/4 × 2. This approach is used by some teachers (Ma 1999). We can use the same principle with fractions. Because 1/2 = 1 ÷ 2, we can write 1 3/4 ÷ 1/2 = 1 3/4 ÷ (1 ÷ 2). This is equal to (1 3/4 ÷ 1) × 2 = 1 3/4 × 2.

Students can use proportional reasoning in division. After solving a sequence of problems like 4 ÷ 4 = 1, 8 ÷ 4 = 2 and 16 ÷ 4 = 4 where the divisor is constant, students will notice that as the dividend increases by a factor of two so does the result. They can realize that the next problem 32 ÷ 4 has to be two times bigger (direct proportionality). Likewise, from looking at a sequence of problems like 8 ÷ 4 = 2, 8 ÷ 2 = 4, 8 ÷ 1 = 8, they can see that the dividend is constant and as the divisor becomes half as great, each result is twice the size as the previous, so that 8 ÷ 1/2 = 16 (inverse proportionality).

Multiplication of Fractions

A sound understanding of multiplication of fractions is also necessary. One meaning of multiplication of fractions arises in situations such as 1/2 of 1/4. It is important to make the connection of symbol to language and also to know some of the common sources of misunderstanding. Whereas "4 divided in two" is written as 4 ÷ 2, "4 divided in half" actually means the same as "1/2 of 4", that is, 1/2 × 4. The statement 4 ÷ 1/2 is read as 4 divided by 1/2. Students also need to understand the area model for multiplication of fractions. If a rectangle has sides that correspond to fractions, its area will be the product of its sides (see fig. 25.3). The area model can also provide a rationale for the procedure "multiply across."

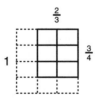

Fig. 25.3. The product 2/3 × 3/4

Compositions of Operations

Students need to be able to transform combinations of operations, for example dividing by a quotient of whole numbers, such as $16 \div (8 \div 2) = (16 \div 8) \times 2$. It is important that students learn to deal with a division as a mathematical object, not only as a computation to be done. This is different from the common way to teach about parenthesis: "Compute whatever is in parenthesis first." Other combinations of operations are multiplying by a division, such as $4 \times (6 \div 3) = (4 \times 6) \div 3$, and dividing by a product, such as $16 \div (4 \times 2) = (16 \div 4) \div 2$.

Division of fractions is also equivalent to a combination of operations. Some students discover this on their own. Pirie (1988) reports that in a class with students aged 11–12 working with pictorial representations, one girl saw that dividing by a fraction was the same as multiplying by the denominator and then dividing by the numerator. We can express this in general as $a \div b/c = (a \times c) \div b$. Another way to look at a division by a fraction as a composition of operations is to first divide by the numerator, and then multiply by the denominator, $a \div b/c = (a \div b) \times c$.

Connections to Algebra

Several aspects that can be stressed in the division of fractions have been identified to help the transition from arithmetic to algebra (Kieran 1990):

- make explicit the procedures to solve arithmetic problems;
- use the equal sign to express a symmetric and transitive relation, rather than just to announce a result;
- consider chains of numbers and operations as mathematical objects and not only as procedures to obtain an answer; and
- pay attention to the method or process and not only to the answer.

Algebraic techniques can help students see analytical proofs from a different perspective. The use of a variable can emphasize that $1/2 \div 1/3$ is a mathematical object in its own right, $1/2 \div 1/3 = x$. We can also write the relationship as $x \times 1/3 = 1/2$, and use algebraic procedures to solve equations to see that $x = 1/2 \times 3$. Students can also use algebraic notation to write in general terms the rules that they formulated for specific numbers. They can express relations such as $a \div b = a/b$, $a/b \div c/b = a \div c$, or $1 \div a/b = b/a$.

Students who understand the relationship between division and multiplication, and the role of reciprocals can use different approaches to solve equations. They can think of a problem either as division by a number or as multiplication by the reciprocal, depending on which makes it easier to solve.

Division of fractions is not the only instance where an operation (division) is changed by the inverse operation (multiplication) at the same time that the corresponding number is changed by its inverse (reciprocal). When working with integers, for example, instead of subtracting a number we can add its (additive) inverse, $a - b = a + (-b)$.

CONCLUSION

As has been documented by researchers, it is not easy for teachers to develop profound understanding of division of fractions and establish the necessary connections on their own. Furthermore, in the United States, division of fractions has been taught for so long by giving a procedure with no explanation why it works that some of today's teachers and prospective teachers only know that method. In order to break the cycle of superficial knowledge of procedures, the kind of knowledge described in this chapter needs to become part of the systematic preparation teachers receive in both preservice and in-service courses.

REFERENCES

Armstrong, Barbara, and Nadine Bezuk. "Multiplication and Division of Fractions: The Search for Meaning." In *Providing a Foundation for Teaching Mathematics in the Middle Grades*, edited by Judith T. Sowder and Bonnie Schappelle, pp. 85–119. Albany, N.Y.: State University of New York Press, 1995.

Kieran, Carolyn. "Cognitive Processes in Learning School Algebra." In *Mathematics and Cognition*, edited by Pearl Nesher and Jeremy Kilpatrick, pp. 96–112. Cambridge, England: Cambridge University Press, 1990.

Kieren, Thomas E. "Rational and Fractional Numbers as Mathematical and Personal Knowledge." In *Analysis of Arithmetic for Mathematics Teaching*, edited by Gaea Leinhardt, Ralph Putnam, and Rosemary A. Hattrup, pp. 323–71. Hillsdale, N.J.: Lawrence Erlbaum Associates, 1992.

Ma, Liping. *Knowing and Teaching Elementary Mathematics*. Mahwah, N.J.: Lawrence Erlbaum Associates, 1999.

Piaget, Jean, Bärbel Inhelder, and Alina Szeminska. *The Child's Conception of Geometry*. New York: Basic Books, 1960.

Pirie, Susan E. B. "Understanding: Instrumental, Relational, Intuitive, Constructed, Formalised … ? How Can We Know?" *For the Learning of Mathematics* 8, no. 3 (1988): 2–6.

Rowland, Tim. "Dividing by Three-Quarters: What Susie Saw." *Mathematics Teaching* 160 (1997): 30–33.

Toluk, Zulbiye. "Children's Conceptualization of the Quotient Subconstruct of Rational Numbers." Ph. D. diss., Arizona State University, 1999.

26

Connecting Informal Thinking and Algorithms:
The Case of Division of Fractions

Daniel Siebert

CHILDREN often lack a ready understanding for operations involving rational numbers, because these operations are frequently equated with seemingly nonsensical algorithms, such as the algorithm for division of fractions. For children, the traditional algorithm for dividing fractions, the invert and multiply (IM) rule, does not seem connected to division in any way. The IM rule has no division sign, nor is it compatible with how children commonly conceive of division as splitting things up evenly or finding how many times one number can be subtracted from another. Children have no reason to believe that "flipping" the divisor and multiplying the resultant fractions together does any dividing whatsoever.

Of course, we as teachers can try to convince children that the IM rule really is division, but if we use traditional approaches to teach the IM rule, we will fail miserably. Symbolic proofs that the IM rule works for fraction division do not help children see the IM rule as division, because the procedures invoked to prove the IM rule are every bit as meaningless to children as the IM rule itself. Empirical arguments—our attempts to show children that we get the same answer for simple division problems regardless of whether we solve them with pictures, manipulatives, or the IM rule—do not go far enough in establishing meaning for division. Unless we can actually point to where we "invert and multiply" in our pictures, children will still see the IM rule as an unexplainable and mysterious short cut to fraction division.

Just because traditional approaches to teaching division of fractions do not sufficiently justify and explain the IM rule does not mean that fraction division using the IM rule has to remain an unfathomable mystery. Children can link their informal thinking about division, that dividing means splitting something up evenly or computing how many times one number can be subtracted from another, to fraction division and the IM rule in a way that they can see the IM rule as division. In other words, by starting their study of the division of fractions with their informal thinking about the two basic types of division situations, children can discover ways to draw pictures for fraction division in which they can actually see what it means to invert and multiply.

Two Meanings for Whole-Number Division

For teachers to help children develop meaningful conceptions of division of fractions, they must first clearly understand whole-number division. As noted above, there are two common meanings for whole-number division: finding how many times one number can be subtracted from another, or splitting something up into equal groups. Mathematics educators refer to the first interpretation of division as measurement (or quotitive), and the second as sharing (or partitive). These two interpretations represent significantly different views of division and arise from two different types of division situations, illustrated in these examples:

Problem 1: Tony has eight pieces of candy. If he plans to give his friends two pieces of candy each, to how many friends can he give candy?

Problem 2: Tony has eight pieces of candy. If he has two friends to whom he wants to give candy, how many pieces should he give to each person so that each friend gets the same amount of candy?

Both these problems can be solved by dividing 8 by 2. However, different activities are needed to model the solutions to these two problems. We solve Problem 1 by forming groups of 2 candies to determine how many groups can be made. This is a measurement situation, because we are measuring how many groups of 2 are in 8. Our answer, 4, represents the number of 2-candy groups that we made (or could take away) from 8 candies. In general, measurement situations involve finding how many groups can be made when the total amount and amount per group are known:

$$\left(\begin{array}{c} Total \\ amount \end{array}\right) \div \left(\begin{array}{c} Amount \\ per\,group \end{array}\right) = \left(\begin{array}{c} Number \\ of\,groups \end{array}\right)$$

In contrast, we solve Problem 2 by partitioning the candies into two equal groups to determine how much each person gets. This is a sharing situation, because we are sharing 8 between 2 groups. Our answer, 4, represents the number of candies that are in each of the 2 groups. In general, sharing situations involve finding out how much is in each group when the total amount and number of groups are known:

$$\left(\begin{array}{c}\textit{Total} \\ \textit{amount}\end{array}\right) \div \left(\begin{array}{c}\textit{Number} \\ \textit{of groups}\end{array}\right) = \left(\begin{array}{c}\textit{Amount} \\ \textit{per group}\end{array}\right)$$

Because there are two interpretations for division, we can think of any number sentence involving whole-number division in two different ways, either as measurement or sharing. For example, we can think of 15 ÷ 3 = 5 as a measurement situation by asking the question, "How many groups of 3 are in 15?" Or we can create a sharing situation by asking, "If 15 were split equally into three groups, how much would be in each group?" The distinction between measurement and sharing is important, because when these interpretations are extended to the division of fractions, they yield fundamentally different ways of conceptualizing the IM rule.

THE IM RULE FROM A MEASUREMENT PERSPECTIVE

The measurement interpretation of the division of whole numbers can be extended directly to the division of fractions without modification. Consider, for example, the division problem 1 1/2 ÷ 3/5. Based on the measurement interpretation of division, this problem is equivalent to asking how many groups of 3/5 are in 1 1/2. One solution with pictures is found in figure 26.1. To create this picture, draw 1 1/2 copies of a rectangle to represent the dividend in the division problem. To see how many 3/5 of a rectangle can be made from 1 1/2 rectangles, partition the existing rectangles into fifths, and then group the one-fifth sections into sets of three, corresponding to groups of 3/5. Two complete groups of 3/5 can be formed. The remainder appears to be half of another 3/5. A smaller partition into tenths verifies that the remainder contains 3 of the 6 one-tenths that are needed to make another 3/5. Thus, the remainder is indeed half of another 3/5. Consequently, there are 2 1/2 groups of 3/5 in 1 1/2. This means that 1 1/2 ÷ 3/5 = 2 1/2.

Although figure 26.1 provides rich meaning for fraction division from a measurement perspective, it does not show why the IM rule works. We can justify the IM rule for 1 1/2 ÷ 3/5, however, by first developing a method for seeing the reciprocal of 3/5 in our picture. Consider the division problem 1 ÷ 3/5, which is equivalent to asking how many groups of 3/5 are in 1. In figure 26.1, we can see that the fraction 3/5 is composed of three 1/5s. Because it takes three 1/5's to make 3/5, each of these 1/5 pieces represents 1/3 of 3/5.

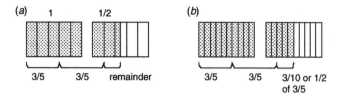

Fig. 26.1. Picture solution to 1 1/2 ÷ 3/5 based on the measurement interpretation of division: (*a*) Draw 1 1/2 and partition each large rectangle into fifths, which makes two groups of 3/5 in 1 1/2, with a little left over. (*b*) Partition the large rectangles into tenths to see that the remainder contains half of the 6 tenths necessary to make another 3/5. Thus, the remainder is a half group of 3/5. Consequently, there are 2 1/2 groups of 3/5 in 1 1/2.

But there are exactly five of these 1/5 pieces in 1. Since each 1/5 represents 1/3 of 3/5, and there are five 1/5's in 1, there are five 1/3's of 3/5 in 1, or 5/3 groups of 3/5 in 1. Thus, 5/3, the reciprocal of 3/5, tells us how many 3/5's are in 1 and can be seen by conceptualizing 1/5 as both 1/5 of 1 and 1/3 of 3/5 (see fig. 26.2).

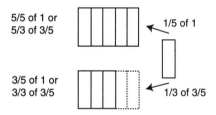

Fig 26.2. The relationship between a fraction and its reciprocal

Once we realize that 5/3 is the number of groups of 3/5 in 1, we can justify the IM rule for 1 1/2 ÷ 3/5. We do this by conceiving of 1 1/2 as being composed of two parts, a whole and another half of a whole, and asking the question "How many 3/5's are in each of these parts?" There are 5/3 groups of 3/5 in the whole and another half of 5/3 groups of 3/5 in half of the whole. Thus, there are 1 1/2 of 5/3 groups of 3/5 in 1 1/2, or 1 1/2 × 5/3 groups of 3/5 in 1 1/2. This shows that 1 1/2 ÷ 3/5 = 1 1/2 × 5/3. In general, because the dividend (1 1/2) represents the number of wholes that can be used to form groups of the divisor (3/5), we can multiply the reciprocal of the divisor (5/3) and the dividend to find out how many of the divisor are in that many wholes.

THE IM RULE FROM A SHARING PERSPECTIVE

At first glance, the sharing interpretation of division does not seem to be a viable way to interpret many division-of-fractions number sentences, particularly for problems like 1/2 ÷ 2/3. We can no longer write a story problem about sharing candy for this division problem, because how can 1/2 a piece of candy be shared with 2/3 of a person? However, there are ways to interpret 1/2 ÷ 2/3 that do not seem to fall under the measurement interpretation, such as the following story problem:

> Melissa used 2/3 of a can of frosting to frost 1/2 a cake. How much of the cake could she frost with the whole can of frosting?

One possible solution to this problem is to realize that if 2/3 of a can frosts 1/2 a cake, then 1/3 of a can frosts 1/4 a cake. There are three 1/3's of a can in a whole can of frosting, and Melissa can frost 1/4 of a cake with each 1/3 of a can. Therefore, Melissa should be able to frost 3/4 of the cake with a whole can of frosting. Interestingly, we can get the same answer of 3/4 by dividing 1/2 by 2/3. Is this just a coincidence? Certainly we were not using the measurement interpretation of division to solve the problem, because we never considered the question of how many groups of 2/3 are in 1/2. Instead, we were conceiving of 1/2 as being 2/3 of something we wanted to find. In other words, we were trying to solve the problem 2/3 × ? = 1/2. To solve this problem using algebra, we might divide both sides by 2/3, which suggests that perhaps our reasoning was indeed compatible with the operation of division.

If we assume for a moment that we really can solve the cake problem by dividing, then we might represent the division problem as follows:

$$\left(\begin{array}{c}\textit{Total amount}\\ \textit{of frosted cake}\end{array}\right) \div \left(\begin{array}{c}\textit{Fraction}\\ \textit{of a can}\end{array}\right) = \left(\begin{array}{c}\textit{Amount of frosted cake}\\ \textit{for 1 can of frosting}\end{array}\right)$$

Compare this with the way we conceptualized division as sharing:

$$\left(\begin{array}{c}\textit{Total}\\ \textit{amount}\end{array}\right) \div \left(\begin{array}{c}\textit{Number}\\ \textit{of groups}\end{array}\right) = \left(\begin{array}{c}\textit{Amount}\\ \textit{per group}\end{array}\right)$$

Note that the goal of both actions seems to be the same: to determine how much of the dividend should be associated with 1 unit of the divisor. In the candies situation before, we split 8 candies among 2 people to find out how many candies a single person received. In the cake situation, we engaged in proportional reasoning to determine how much cake 1 can of frosting could cover. This suggests that the sharing interpretation of division does apply to the division of fractions, with some modifications. Instead of relying on the action of divvying out the total amount to a whole number of groups or units, we use the actions of sharing and duplicating to determine how much a whole group or unit would receive.

To illustrate how the sharing interpretation can be extended to fraction division, consider the problem 1 1/2 ÷ 3/5, which we solved earlier using a measurement interpretation. We can conceive of this problem as a sharing situation by asking the guiding question, "If 3/5 of a group gets 1 1/2, then how much should a whole group get?" This creates a sharing situation because it matches the way we conceptualized whole-number division from a sharing perspective. Just like sharing for whole numbers, the dividend (1 1/2) and the divisor (3/5) correspond to the total amount and the number of groups, respectively. The answer to 1 1/2 ÷ 3/5 tells us how much a single group gets, which also matches the meaning for the quotient in a whole-number sharing situation.

What remains different between sharing for whole numbers and fractions is the way we conceive of the solution. The solution to 1 1/2 ÷ 3/5 from a sharing perspective not only involves the action of sharing, but of duplicating as well, which can be seen in the following reasoning. If 3/5 of a group gets 1 1/2, then 1/5 of a group, which is 1/3 of 3/5, would get 1/3 of 1 1/2, which is 1/2 (see fig. 26.3a). This reasoning is compatible with the action of sharing the 1 1/2 between the three 1/5's so that each 1/5 gets 1/3 of 1 1/2, and corresponds to dividing 1 1/2 by 3. One-half, however, is not the answer to 1 1/2 ÷ 3/5, because it is not the amount that a whole group gets. To find how much a whole group gets, we take one of the 1/5's of a group and duplicate it four times to create a whole group. One-half, the corresponding amount that 1/5 of a group gets, is also duplicated four times to yield five 1/2's, or 5/2, the amount a whole group gets (see fig. 26.3b). This duplicating process is equivalent to multiplying 1/2 by 5 to get 2 1/2 for the whole group. In summary, to find how much a whole group received, we divided 1 1/2 by 3 and then multiplied by 5.

The solution to 1 1/2 ÷ 3/5 from a sharing perspective suggests an alternative meaning for 5/3, the reciprocal of 3/5. Rather than viewing 5/3 as the number of 3/5's in 1, as we did with the measurement interpretation, we can conceive of it as an operator that first shrinks a quantity to a third of its original size and then expands the shrunken quantity to five times its shrunken size. This process of shrinking and expanding matches the results of sharing and duplicating in our solution above. We shrank both the 1 1/2 and 3/5 to 1/3 of their size when we shared the 1 1/2 across the 3/5. Then we increased both quantities to five times their size when we duplicated them. In other

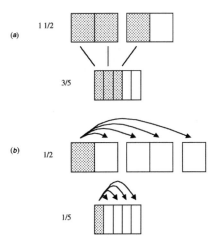

Fig. 26.3. Picture solution for the sharing interpretation of 1 1/2 ÷ 3/5:
(a) First, share 1 1/2 between the 3/5 to find that each 1/5 gets 1/2. (b)
To find how much a whole gets, duplicate the 1/5 four times to produce
a whole. The 1/2, the amount 1/5 gets, is also duplicated four times to
produce 2 1/2, the amount a whole gets.

words, to find out how much a whole group would receive if 3/5 of a group
received 1 1/2, we took 5/3 of 3/5 and 1 1/2, which told us that a whole
group received 2 1/2. The reciprocal of the divisor, when conceived as an
operator, is actually the operator necessary to change the divisor into a 1.
Because of the proportional relationship that exists between the divisor and
dividend, we must also apply the operator to the dividend, or in other
words, multiply the dividend by the reciprocal of the divisor. This is in fact
the IM rule.

A COMPARISON OF THE TWO INTERPRETATIONS OF THE IM RULE

As demonstrated above, the IM rule can be derived from our images of
what it means to divide fractions. In fact, because we can think of fraction
division in two different ways, we can also conceive of the IM rule in two different ways. Table 26.1 summarizes these two ways of conceptualizing division of fractions and the IM rule for the division problem 2 1/2 ÷ 3/4.

TABLE 26.1

Summary of the measurement and sharing interpretations for division of fractions

	Measurement	Sharing
Situations	Joel is walking around a circular path in a park that is 3/4 miles long. If he walks 2 1/2 miles before he rests, how many times around the path did he travel?	Joel is walking around a circular path in a park. If he can walk 2 1/2 miles in 3/4 of an hour, how far can he walk in an hour, assuming he walks at the same speed?
Guiding question for interpreting 2 1/2 ÷ 3/4	How many groups of 3/4 are in 2 1/2?	If 3/4 of a group gets 2 1/2, how much does a whole group get?
Meaning of reciprocal	The reciprocal 4/3 means there are 4/3 groups of 3/4 in 1.	The reciprocal 4/3 is the operator necessary to shrink 3/4 to 1/4 and then expand 1/4 to 1.
Reason for multiplying the dividend by the reciprocal of the divisor	There are 4/3 groups of 3/4 in 1. There are 2 1/2 times as many groups of 3/4 in 2 1/2 as there are in 1. Thus, there are 2 1/2 × 4/3 groups of 3/4 in 2 1/2.	Since we shrink/expand 3/4 by 4/3 to get 1 whole group, we have to shrink/expand 2 1/2 by 3/4 in order to find out how much the whole group gets.

TEACHING THE TWO INTERPRETATIONS

Children can develop meaningful images for the division of fractions by reasoning about real-world contexts that involve fraction division and making connections between their solutions to these problems and their understanding of whole-number division. For example, to develop an understanding of fraction division from a sharing perspective, children can begin by first exploring situations similar to the cake problem. After they become comfortable with their images of sharing and duplicating, they can talk about how their actions and images are similar to, and different from, whole-number division problems involving sharing. Through these discussions, children can extend their understanding of division to situations involving fractions.

Once children possess meaningful images for fraction division, they are then able to discover and find meaning for the IM rule. Teachers can help them do this by focusing their attention on what a reciprocal means. For the measurement interpretation, teachers can give children problems that involve a dividend of 1 so that children can see that the reciprocal of the divisor, the answer to the division problem, is how many of the divisor are in 1. Once children understand what the reciprocal means, the teacher can then change the dividend to another number (leaving the divisor the same) and ask the children to use their understanding of the reciprocal to solve the new problem. This will help children think about why one might multiply the reciprocal times the dividend.

To see why the IM rule works from a sharing perspective, children must link their sharing and duplicating actions for division with their images for multiplication by a fraction. They can do this by first creating picture solutions to carefully chosen division and multiplication problems, and then comparing their solutions. For example, teachers can ask children to draw pictures for 1 1/2 ÷ 3/5 and 5/3 × 1 1/2. If children are encouraged to use a sharing perspective to solve the division problem, then their pictures to both of these two problems will be similar to the picture in figure 26.3. Teachers can engage children in conversations about the similarities between the solutions to these two problems to help children form connections between division from a sharing perspective and multiplication by the reciprocal of the divisor.

Children need to learn to use both interpretations of fraction division, because they will encounter both sharing and measurement situations involving fractions. Measurement situations may be the easiest for children to understand, given the close similarities between the measurement interpretation for whole-number and fraction division. Thus, measurement situations may be the best place for teachers to begin their instruction of fraction division. However, teaching only the measurement interpretation is insufficient. For children to be able to quantify intensive quantities such as speed and slope, they must also understand the sharing interpretation of fraction division, because intensive quantities arise from sharing situations. Once children connect their images of sharing and measurement to the IM rule, the IM rule can become a meaningful tool that they can use to solve a wide range of interesting and important division problems.

Classroom Challenge

Susan Lamon

Marquette University, Milwaukee, Wisconsin

Fraction Problem

Find three fractions between 7/11 and 7/12.

$$\text{Annie}$$

$$\frac{7}{12} \qquad\qquad\qquad \frac{7}{11}$$

$$\frac{7}{11\frac{1}{2}} \qquad \frac{7}{11\frac{1}{4}} \qquad \frac{7}{11\frac{1}{5}}$$

$$\frac{7}{\frac{23}{2}} \qquad \frac{7}{\frac{45}{4}} \qquad \frac{7}{\frac{56}{5}}$$

$$\frac{14}{23} \qquad \frac{28}{45} \qquad \frac{35}{56}$$

Annie, a sixth grader, was comfortable with complex fractions. She explained, "You can get fractions in between those two by making the bottom numbers in between 11 and 12. You can make any number of fractions you want between them." Her teacher wondered if she was sure that those numbers were between 7/11 and 7/12 and asked her to show it. "OK," Annie said. "7/12 = 0.583 and 7/11 = 0.636. so let's pick 7/(11 1/2). If you double the top and the bottom, you get 14/23, and that's 0.068, so it's in there.

Classroom Challenge

Susan Lamon

Marquette University, Milwaukee, Wisconsin

Fraction Problem

Find three fractions between 1/8 and 1/9.

Martin

$$\frac{1}{9} \qquad \frac{8}{8} \div 9 \qquad\qquad \frac{1}{8} = \frac{9}{8} \div 9$$

$$1\frac{1}{8} \div 9$$

$$1\frac{1}{9} \qquad 1\frac{1}{10} \qquad 1\frac{1}{11}$$

$$\frac{10}{9} \qquad \frac{11}{10} \qquad \frac{12}{11}$$

$$\frac{1}{9} \qquad \boxed{\frac{10}{81} \qquad \frac{11}{90} \qquad \frac{12}{99}} \qquad \frac{1}{8}$$

Martin was just finishing fifth grade when he answered this challenge question. He insisted that his teacher specify whether the fractions were to be equally spaced or not, since he knew how to do it both ways. This is his method for the case where spacing is not important. Martin first rewrote the fracton 1/8 so that it had 9 in the denominator. He did this by noticing that $1/9 = 1 \div 9$ and $1/8 = 9/8 \div 9$. Now he was ready to form new fractions between 1/9 and (1 1/8)/9 . All the denominators would be 9. To form the numerators, he stayed between 1 and (1 1/8) by adding to 1 fractions smaller than 1/8 (for example, 1 + 1/9, 1 + 1/10, 1 + 1/11, and so on.)

Classroom Challenge

Judy Wells

University of Southern Indiana

Carroll Wells

Western Kentucky University

A Fraction Activity

1. Draw a large circle and mark its center. Fold the circle in half. What is the creased line called? Fold in half again to determine the true center of the circle. What angle has been formed? How many degrees are in a circle? What is the distance from the center to the circles' circumference called?

2. Mark a point on the circumference of the circle. Fold the point to the center. What is the new segment called? What is the part of the circumference called?

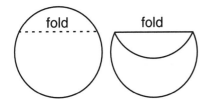

3. Fold again to the center, using one endpoint of the chord as an endpoint for the new chord.

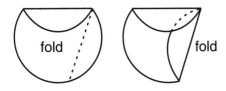

4. Fold the remaining arc to the center. Compare your equilateral triangle with that of your neighbor. Throughout the rest of the activity define the area of your triangle as one unit.

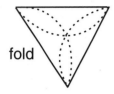

5. Find the midpoint of one of the sides of your triangle. Fold the opposite vertex to the midpoint. What is the area of the isosceles trapezoid if the area of the original triangle is one unit?

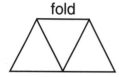

6. Note that the trapezoid consist of three congruent triangles. Fold one of these triangles over the top of the middle triangle. What polygon has been formed? What is its area?

7. Fold the remaining triangle over the top of the other two triangles. What shape is formed? What is its area? This triangle is similar to the original triangle.

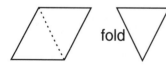

8. Place the three folded-over triangles in the palm of your hand, and open it up to form a three-dimensional shape. What new polyhedron has been formed? What is its surface area?

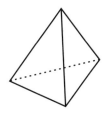

9. Unfold the polyhedron so that the large equilateral triangle is again considered. Fold each vertex to the center of the circle. What is the area of the resulting hexagon?

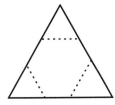 

10. Turn the hexagon over and push gently so that the hexagon folds up to form a truncated tetrahedron.

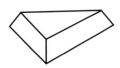

11. Using only the fold lines already determined, create different polygons and determine their area. Using only the existing fold lines, construct polygons with the following areas. Name each polygon and draw a sketch.

1/4 unit   1/2 unit   19/36 unit

2/3 unit   3/4 unit   7/9 unit

8/9 unit   7/18 unit   23/36 unit